Lecture Notes in Mathematics

Edited by J.-M. Morel, F. Takens and B. Teissier

Editorial Policy
for the publication of monographs

1. Lecture Notes aim to report new developments in all areas of mathematics – quickly, informally and at a high level. Monograph manuscripts should be reasonably self-contained and rounded off. Thus they may, and often will, present not only results of the author but also related work by other people. They may be based on specialized lecture courses. Furthermore, the manuscripts should provide sufficient motivation, examples and applications. This clearly distinguishes Lecture Notes from journal articles or technical reports which normally are very concise. Articles intended for a journal but too long to be accepted by most journals, usually do not have this "lecture notes" character. For similar reasons it is unusual for doctoral theses to be accepted for the Lecture Notes series.

2. Manuscripts should be submitted (preferably in duplicate) either to one of the series editors or to Springer-Verlag, Heidelberg. In general, manuscripts will be sent out to 2 external referees for evaluation. If a decision cannot yet be reached on the basis of the first 2 reports, further referees may be contacted: the author will be informed of this. A final decision to publish can be made only on the basis of the complete manuscript, however a refereeing process leading to a preliminary decision can be based on a pre-final or incomplete manuscript. The strict minimum amount of material that will be considered should include a detailed outline describing the planned contents of each chapter, a bibliography and several sample chapters.
Authors should be aware that incomplete or insufficiently close to final manuscripts almost always result in longer refereeing times and nevertheless unclear referees' recommendations, making further refereeing of a final draft necessary.
Authors should also be aware that parallel submission of their manuscript to another publisher while under consideration for LNM will in general lead to immediate rejection.

3. Manuscripts should in general be submitted in English.
Final manuscripts should contain at least 100 pages of mathematical text and should include
- a table of contents;
- an informative introduction, with adequate motivation and perhaps some historical remarks: it should be accessible to a reader not intimately familiar with the topic treated;
- a subject index: as a rule this is genuinely helpful for the reader.

Continued on inside back-cover

Lecture Notes in Mathematics 1767

Springer
Berlin
Heidelberg
New York
Barcelona
Hong Kong
London
Milan
Paris
Tokyo

Jie Xiao

Holomorphic Q Classes

Springer

Author

Jie XIAO
Department of Mathematics and Statistics
Concordia University
1455 de Maisonneuve Blvd. West
H3G 1M8 Montreal, Quebec, Canada

E-mail: jxiao@mathstat.concordia.ca

Cataloging-in-Publication Data applied for

Die Deutsche Bibliothek - CIP-Einheitsaufnahme

Xiao, Jie:
Holomorphic Q classes / Jie Xiao. - Berlin ; Heidelberg ; New York ;
Barcelona ; Hong Kong ; London ; Milan ; Paris ; Tokyo : Springer, 2001
(Lecture notes in mathematics ; 1767)
ISBN 3-540-42625-6

Mathematics Subject Classification (2000):
30D55, 30H05, 31A20, 32A37, 41A15, 46E15, 46G10, 47B33, 47B38

ISSN 0075-8434
ISBN 3-540-42625-6 Springer-Verlag Berlin Heidelberg New York

Springer-Verlag Berlin Heidelberg New York
a member of BertelsmannSpringer Science+Business Media GmbH

http://www.springer.de

Typesetting: Camera-ready $\TeX$ output by the author

SPIN: 10852603 41/3142-543210/du - Printed on acid-free paper

Preface

One of the fundamental problems in any area of mathematics is to determine the distinct variants of an object under consideration. As for complex-functional analysis, one is interested, for example, in studying the equivalent representations of the conformally invariant classes of holomorphic functions. This problem is addressed here for the holomorphic Q classes.

For $p \in [0, \infty)$ and dm – the element of the two dimensional Lebesgue measure, we say that f, a holomorphic function in the unit disk $\mathbf{D}$, is of the class $\mathcal{Q}_p$ provided

$$E_p(f) = \sup\left\{\left(\int_{\mathbf{D}} |f'(z)|^2 \left(\log\left|\frac{1-\bar{w}z}{w-z}\right|\right)^p dm(z)\right)^{1/2} : w \in \mathbf{D}\right\} < \infty.$$

It is clear that each $\mathcal{Q}_p$ can serve as a sample of the conformally invariant classes in the sense of: $E_p(f \circ \sigma) = E_p(f)$ for all $f \in \mathcal{Q}_p$ and $\sigma \in Aut(\mathbf{D})$ – the group of all conformal automorphisms of $\mathbf{D}$.

The goal of this monograph is to bring the major features of $\mathcal{Q}_p$ to light, in particular, to characterize $\mathcal{Q}_p$ in different terms. More precisely:

Chapter 1 contains some (most) basic properties of $\mathcal{Q}_p$ such as Möbius boundedness, image area and higher derivatives. The aim is to show that the classes $\mathcal{Q}_p$, $p \in (0, 1)$ are of independent interest.

Chapter 2 discusses the problem of how a $\mathcal{Q}_p$ can be embedded into a Bloch-type space and vice versa. This problem will be solved by considering boundedness and compactness of a composition operator acting between two spaces.

Chapter 3 describes the coefficients of either Taylor or random series of functions from $\mathcal{Q}_p$. In particular, we will see that there is a big difference between the cases: $p \in (0, 1)$ and $p = 1$.

Chapter 4 exhibits a geometric way to understand $\mathcal{Q}_p$, that is, p-Carleson measure characterization of $\mathcal{Q}_p$. This simple but important property induces certain deep relations between $\mathcal{Q}_p$ and the mean Lipschitz spaces, as well as the Besov spaces which are conformally invariant, too.

Chapter 5 characterizes the inner and outer functions in $\mathcal{Q}_p$ by means of p-Carleson measure and other two conformally invariant measures: Poisson measure and hyperbolic measure.

Chapter 6 gives the boundary value behavior of a $\mathcal{Q}_p$-function for $p \in (0, 1)$. This allows us to study $\mathcal{Q}_p$ via those non-holomorphic functions on the unit circle $\mathbf{T}$ and even on the exterior of the unit disk $\mathbf{C} \setminus \mathbf{D}$, and hence leads to a consideration of harmonic analysis.

Chapter 7 explores a list of properties of $\mathcal{Q}_p(\mathbf{T})$ (i.e. Q class on $\mathbf{T}$). Specially, the $\mathcal{Q}_p(\mathbf{T})$-solutions of the $\bar{\partial}$-equation produce a decomposition of $\mathcal{Q}_p$ through the bounded functions on $\mathbf{T}$. As applications, the corona theorem and interpolation theorem related to $\mathcal{Q}_p$ are established.

Chapter 8 deals with a localization of $\mathcal{Q}_p(\mathbf{T})$ based on the dyadic partitions of all subarcs of $\mathbf{T}$. The results enable us to recognize $\mathcal{Q}_p$ from mean oscillation to dyadic model, and finally to wavelet basis.

The exposition is at as elementary a level as possible, and it is intended to be accessible to graduate students with a basic knowledge of complex-functional-real analysis. The material of this monograph has been collected from a series of talks that I gave over the past six years most in Canada, China, Finland, Germany, Greece and Sweden, but also from a lecture course at University of La Laguna in the fall semester of 1999. The selection of topics is rather arbitrary, but reflects the author's preference for the analytic approach. There is no attempt to cover all recent advances (for instance, Q classes of higher dimensions), and yet, it is hoped that the reader will be intrigued by this monograph and will, at some point, read the notes presented at the end of each chapter as well as the papers listed in the references, and proceed to a further research.

Here, I owe a great debt of gratitude to the many people who assisted me with this work. R. Aulaskari and M. Essén read the whole manuscript, catched a number of errors and offered many helpful suggestions. G. Dafni, P. Gauthier and K. J. Wirths read parts of the manuscript and contributed significantly, by valuable queries and comments, to the accuracy of the final version. M. Anderson, S. Axler, H. Carlsson, D. C. Chang, Y. He, S. Janson, H. Jarchow, F. Jafari, H. Kisilevsky, L. Lindahl, A. Nicolau, J. Peetre, L. Peng, F. Pérez-González, H. Proppe, S. Ruscheweyh, W. Sander, A. Siskakis, W. Smith, K. Sten, D. Stegenga, M. Wong, K. Xiong and G. Zhang made friendly advice and warm encouragement.

The following also gave aid and comfort. The Alexander von Humboldt Foundation, Germany and the Swedish Institute, Sweden supported my work on this book. The Institute of Mathematics at Technical University of Braunschweig supplied the computer facilities, and moreover its faculty member H. Weiss kindly helped me create LaTeX working directory and taught me much knowledge about computer. The Department of Mathematics and Statistics at Concordia University provided a good place to carry my writing and revising through to the end. Without their help I would not have gone ahead with publishing this book.

I am grateful to the editors of Springer-Verlag for accepting this monograph for publication in the LNM series, as well as to S. Zoeller for the efficient handling of the editing.

Finally, I want to express my special thanks to my wife, Xianli, and my son, Sa, for their understanding and support.

Montreal, July 2001 *Jie Xiao*

Contents

1. Fundamental Material

In this chapter we present the major notational conventions and the fundamental facts on the holomorphic Q classes (such as definition, inclusion, image area and higher derivatives) which will be used throughout the monograph.

1.1 Introduction

Here, and elsewhere in this monograph, let $\mathbf{D}$ be the unit disk of the complex plane $\mathbf{C}$, $\mathbf{T}$ its boundary, and dm the element of the Lebesgue area measure on $\mathbf{D}$. The symbol $\mathcal{H}$ is employed to represent the class of holomorphic functions on $\mathbf{D}$ endowed with the topology of uniform convergence on compact subsets of $\mathbf{D}$.

Let $Aut(\mathbf{D})$ be the group of all conformal automorphisms of $\mathbf{D}$, namely, all Möbius transformations of the form $\sigma = \zeta\sigma_w$, where $\zeta \in \mathbf{T}$ and $\sigma_w(z) = (w-z)/(1-\bar{w}z)$ is the symmetry interchanging 0 and $w \in \mathbf{D}$. Denote by $g(z,w) = -\log|\sigma_w(z)|$ the Green function of $\mathbf{D}$ with pole at $w \in \mathbf{D}$. The pseudo-hyperbolic disk of (pseudo-hyperbolic) center $w \in \mathbf{D}$ and (pseudo-hyperbolic) radius $r \in (0,1]$ will be denoted by $\mathbf{D}(w,r) = \{z \in \mathbf{D} : |\sigma_w(z)| < r\}$. Furthermore, with $p \in [0,\infty)$, $f \in \mathcal{H}$ and $w \in \mathbf{D}$ we associate a square root of area integral:

$$E_p(f,w) = \left(\int_{\mathbf{D}} |f'(z)|^2 (g(z,w))^p dm(z)\right)^{1/2}.$$

Definition 1.1.1. *For $p \in [0,\infty)$, a function f in $\mathcal{H}$ is said to be of the class $\mathcal{Q}_p$ in case*

$$E_p(f) = \sup_{w\in\mathbf{D}} E_p(f,w) < \infty.$$

Since the Green function is conformally invariant: $g(\sigma(z),\sigma(w)) = g(z,w)$ for all $\sigma \in Aut(\mathbf{D})$ and $z,w \in \mathbf{D}$, it is nearly obvious that each class $\mathcal{Q}_p$ is conformally invariant in the following sense: if $f \in \mathcal{Q}_p$ then $E_p(f\circ\sigma) = E_p(f)$ for all $\sigma \in Aut(\mathbf{D})$.

The definition of $\mathcal{Q}_p$ comes essentially from the research of three conformally invariant classes: the Dirichlet, BMOA and the Bloch spaces.

• $\mathcal{Q}_0$ can be identified with the Dirichlet space $\mathcal{D}$ (cf. [11]) of $f \in \mathcal{H}$ having symmetric boundary behavior on $\mathbf{T}$:

$$\|f\|_{\mathcal{D}} = \left(\int_{\mathbf{T}}\int_{\mathbf{T}}\left|\frac{f(\zeta)-f(\eta)}{\zeta-\eta}\right|^2 |d\zeta||d\eta|\right)^{1/2} < \infty.$$

• $\mathcal{Q}_1$ coincides with BMOA (cf. [30], [99]), the class of analytic (i.e., holomorphic) functions of bounded mean oscillation on $\mathbf{T}$. Here and afterwards, we say that an $L^2(\mathbf{T})$ function f is of BMO, simply, $f \in BMO(\mathbf{T})$ if

$$\|f\|_{BMO} = \sup_{I\subseteq\mathbf{T}}\left(|I|^{-1}\int_I \left|f(\zeta)-f_I\right|^2 |d\zeta|\right)^{1/2} < \infty,$$

where the supremum is taken over all subarcs I of $\mathbf{T}$ with the arclength $|I|$ (i.e., $\int_I |d\zeta|$), and $f_I = |I|^{-1}\int_I f(\eta)|d\eta|$.

• $\mathcal{Q}_2$ is equal to the Bloch space $\mathcal{B}$ (cf. [133]) consisting of all $f \in \mathcal{H}$ whose expansions have finite upper bound:

$$\|f\|_{\mathcal{B}} = \sup_{z\in\mathbf{D}}(1-|z|^2)|f'(z)| < \infty.$$

In what follows, we will use the notation $U \approx V$ to denote comparability of the quantities, i.e., there are two positive constants c_1, c_2 satisfying $c_1 V \le U \le c_2 V$. Similarly, we say that $U \succeq V$ or $U \preceq V$ if only the first or second inequality holds.

Also, for $p \in [0,\infty)$ and $w \in \mathbf{D}$ we define another square root of area integral of $f \in \mathcal{H}$ to be

$$F_p(f,w) = \left(\int_{\mathbf{D}} |f'(z)|^2(1-|\sigma_w(z)|^2)^p dm(z)\right)^{1/2}.$$

Upon noticing the growth of the Green function $g(\cdot,\cdot)$, we can obtain a simple but useful equivalent description of $\mathcal{Q}_p$.

Theorem 1.1.1. *Let $p \in [0,\infty)$ and $f \in \mathcal{H}$. Then $f \in \mathcal{Q}_p$ if and only if*

$$|\!|\!|f|\!|\!|_{\mathcal{Q}_p} = |f(0)| + \|f\|_{\mathcal{Q}_p} = |f(0)| + \sup_{w\in\mathbf{D}} F_p(f,w) < \infty.$$

Proof. By the inequalities: $-2\log t \ge 1-t^2$, $t \in (0,1]$; as well as

$$-\log t \le 4(1-t^2), \quad t \in (1/4,1), \tag{1.1}$$

it suffices to show:

$$E_p(f,w) \preceq F_p(f,w), \quad w \in \mathbf{D}. \tag{1.2}$$

Because $\int_{\mathbf{T}} |f'(r\zeta)|^2|d\zeta|$ is a nondecreasing function of $r \in (0,1)$, one has

$$\int_{\mathbf{T}} |f'(\zeta/4)|^2|d\zeta| \preceq \int_{\mathbf{D}\setminus\mathbf{D}(0,1/4)} |f'(z)|^2(1-|z|^2)^p dm(z),$$

which implies

$$\int_{\mathbf{D}(0,1/4)} |f'(z)|^2 \left(\log \frac{1}{|z|}\right)^p dm(z) \preceq (F_p(f,0))^2.$$

This, together with (1.1), leads to $E_p(f,0) \preceq F_p(f,0)$ which applies to $f \circ \sigma_w$, and hence (1.2) follows.

Theorem 1.1.1 and its proof can be used to give a conformally invariant description of $\mathcal{Q}_p$ as the Möbius bounded functions in certain weighted Dirichlet spaces.

For $p \in [0,\infty)$, the weighted Dirichlet space $\mathcal{D}_p$ is the space of all $f \in \mathcal{H}$ satisfying

$$|||f|||_{\mathcal{D}_p} = |f(0)| + \|f\|_{\mathcal{D}_p} = |f(0)| + F_p(f,0) < \infty.$$

Thus, $\mathcal{D}_0$, $\mathcal{D}_1$ and $\mathcal{D}_2$ are the usual Dirichlet, Hardy and Bergman spaces: $\mathcal{D}$, H^2 and L^2_a, respectively.

Corollary 1.1.1. *Let $p \in [0,\infty)$. Then*
(i) *$f \in \mathcal{Q}_p$ if and only if $f \in \mathcal{H}$ and $\sup_{w\in\mathbf{D}} |||f \circ \sigma_w - f(w)|||_{\mathcal{D}_p} < \infty$.*
(ii) *$\mathcal{Q}_p$ is a Banach space with respect to the norm $|||\cdot|||_{\mathcal{Q}_p}$.*

Proof. (i) This follows directly from Theorem 1.1.1.

(ii) It is enough to verify the completeness of $\mathcal{Q}_p$. First of all, if $f \in \mathcal{Q}_p$ then we apply the fact that $|f'|^2$ is subharmonic to get that for $w \in \mathbf{D}$,

$$(F_p(f,w))^2 = (F_p(f \circ \sigma_w, 0))^2 \succeq |(f \circ \sigma_w)'(0)|^2. \tag{1.3}$$

Next, let $\{f_n\}$ be a Cauchy sequence in $\mathcal{Q}_p$. By (1.3), there exits a subsequence $\{f_{n_k}\}$ which converges to some $f \in \mathcal{H}$, uniformly on compact subsets of $\mathbf{D}$. It follows from Fatou's lemma that for every integer $k \geq 1$,

$$|||f - f_{n_k}|||_{\mathcal{Q}_p} \leq \limsup_{n\to\infty} |||f_n - f_{n_k}|||_{\mathcal{Q}_p},$$

which produces $f \in \mathcal{Q}_p$ and $f_{n_k} \to f$ in $\mathcal{Q}_p$.

1.2 Inclusion

In this section, we will clarify the differences between the $\mathcal{Q}_p$ classes, in particular, show that each class $\mathcal{Q}_p$, $p \in (0,1)$, is of independent interest. To see this, we will use $\mathbf{N}$ to denote the set of all positive integers, and say that f is in HG, the Hadamard gap class, if $f(z) = \sum_{k=0}^{\infty} a_k z^{n_k}$ is in $\mathcal{H}$ and there is a constant $c > 1$ such that $n_{k+1}/n_k \geq c$ for all $k \in \mathbf{N}$.

Theorem 1.2.1. *Let $p \in [0,1]$ and $f(z) = \sum_{k=0}^{\infty} a_k z^{n_k}$ be in HG. Then*
(i) *$f \in \mathcal{Q}_p$ if and only if $f \in \mathcal{D}_p$ if and only if*

$$\sum_{k=0}^{\infty} 2^{k(1-p)} \sum_{2^k \leq n_j < 2^{k+1}} |a_j|^2 < \infty. \tag{1.4}$$

(ii) *$f \in \mathcal{B}$ if and only if $\sup_k |a_k| < \infty$.*

Proof. (i) In the case $p = 0$, the result is trivial. Accordingly, let us consider the case $p \in (0, 1]$. On the one hand, suppose (1.4) holds. Let $w \in \mathbf{D}$ and $I_k = \{n \in \mathbf{N} : 2^k \le n < 2^{k+1}\}$ for each $k \in \mathbf{N}$. Then by Hölder's inequality and some elementary estimations, we get

$$\begin{aligned}(F_p(f,w))^2 &\le \int_0^1 \Big(\sum_{k=0}^{\infty} n_k|a_k|r^{n_k-1}\Big)^2 \Big(\int_{\mathbf{T}} (1-|\sigma_w(r\zeta)|^2)^p |d\zeta|\Big) dr \\ &\preceq \int_0^1 \Big(\sum_{k=0}^{\infty} n_k|a_k|r^{n_k-1}\Big)^2 \Big(\int_{\mathbf{T}} (1-|\sigma_w(r\zeta)|^2)|d\zeta|\Big)^p dr \\ &\preceq \int_0^1 \Big(\sum_{k=0}^{\infty} n_k|a_k|r^{n_k-1}\Big)^2 (1-r)^p dr \\ &\preceq \sum_{k=0}^{\infty} 2^{-k(1+p)} \Big(\sum_{n_j\in I_k} n_j|a_j|\Big)^2 \\ &\preceq \sum_{k=0}^{\infty} 2^{k(1-p)} \Big(\sum_{n_j\in I_k} |a_j|\Big)^2 .\end{aligned}$$

Since $f \in HG$, there exists a constant $c > 1$ such that the number of Taylor coefficients a_j (of f) when $n_j \in I_k$ is at most 1 plus the integer part $[\log_c 2]$ of $\log_c 2$. This implies

$$(F_p(f,w))^2 \preceq \sum_{k=0}^{\infty} 2^{k(1-p)} \sum_{n_j\in I_k} |a_j|^2,$$

and so $f \in \mathcal{Q}_p$.

On the other hand, if $f \in \mathcal{Q}_p$ then $f \in \mathcal{D}_p$, and hence

$$\begin{aligned}\|f\|^2_{\mathcal{Q}_p} &\ge \int_0^1 \Big(\int_{\mathbf{T}} \Big|\sum_{k=0}^{\infty} n_k a_k (r\zeta)^{n_k}\Big|^2 |d\zeta|\Big)(1-r^2)^p r dr \\ &\succeq \sum_{k=0}^{\infty} n_k^{1-p}|a_k|^2 \\ &\succeq \sum_{k=0}^{\infty} \sum_{n_j\in I_k} n_j^{1-p}|a_j|^2 \\ &\succeq \sum_{k=0}^{\infty} 2^{k(1-p)} \sum_{n_j\in I_k} |a_j|^2,\end{aligned}$$

so that (1.4) holds.

(ii) Assume $f \in \mathcal{B}$. Then for any $r \in (0, 1)$ one has

$$n_k|a_k| = \frac{1}{2\pi}\Big|\int_{\mathbf{T}} f'(r\zeta)(r\zeta)^{1-n_k}|d\zeta|\Big| \preceq \frac{\|f\|_{\mathcal{B}}}{n_k(1-r)^{n_k-1}}.$$

Choosing $r = 1 - n_k^{-1}$, we read $|a_k| \preceq 1$ for all $k \in \mathbf{N}$. Conversely, suppose $\sup_{k\in\mathbf{N}} |a_k| < \infty$, and $K = K(n) = \max\{k : n_k \le n\}$. Since $f \in HG$, there is a constant $c > 1$ such that

$$\frac{1}{n}\sum_{n_j\le n} n_j = \frac{n_K}{n}\sum_{n=0}^{K-1}\frac{n_{K-n}}{n_K} \le \frac{c}{c-1}.$$

Thus,

$$\frac{|zf'(z)|}{1-|z|} \preceq \Big(\sum_{k=0}^{\infty}|z|^k\Big)\Big(\sum_{j=0}^{\infty}n_j|z|^{n_j}\Big) \preceq \sum_{n=1}^{\infty}\Big(\sum_{n_j\le n}n_j\Big)|z|^n \preceq \frac{|z|}{(1-|z|)^2},$$

which gives $f \in \mathcal{B}$.

Corollary 1.2.1. *Let $p, q \in [0,\infty)$. Then $(\mathcal{Q}_q, \|\cdot\|_{\mathcal{Q}_q})$ nonincreases to $(\mathcal{Q}_p, \|\cdot\|_{\mathcal{Q}_p})$ as $q \searrow p$. Moreover,*

(i) *If $p \in (1,\infty)$, then $\mathcal{Q}_p = \mathcal{B}$.*
(ii) *If $p = 1$, then $\mathcal{Q}_p \ne \mathcal{B}$.*
(iii) *If $0 \le p \ne q \le 1$, then $\mathcal{Q}_p \ne \mathcal{Q}_q$.*

Proof. Theorem 1.1.1 implies that if $0 \le p < q < \infty$, then $\mathcal{Q}_p \subseteq \mathcal{Q}_q$ with $\|\cdot\|_{\mathcal{Q}_q} \le \|\cdot\|_{\mathcal{Q}_p}$. Meanwhile, for $w \in \mathbf{D}$, $r \in (0,1)$ and $f \in \mathcal{Q}_q$, one has

$$F_q(f,w) \ge (1-r^2)^{(q-p)/2}\left(\int_{\mathbf{D}(w,r)} |f'(z)|^2(1-|\sigma_w(z)|^2)^p dm(z)\right)^{1/2},$$

which implies

$$\lim_{q\searrow p} F_q(f,w) \ge \left(\int_{\mathbf{D}(w,r)} |f'(z)|^2(1-|\sigma_w(z)|^2)^p dm(z)\right)^{1/2}.$$

Thus, $\mathcal{Q}_p$ consists of those functions for which there is a constant $C(f) > 0$ (depending on f) with $\|f\|_{\mathcal{Q}_q} \le C(f)$ for all $q > p$ and so, can be viewed as a limit space of $\mathcal{Q}_p$ as $q \searrow p$.

Now, we consider the special cases stated as above. First, let $p \in (1,\infty)$. Then $f \in \mathcal{B}$ gives

$$F_p(f,w) \le \|f\|_{\mathcal{B}}\left(\int_{\mathbf{D}} (1-|z|^2)^{-2}(1-|\sigma_w(z)|)^p dm(z)\right)^{1/2}, \qquad (1.5)$$

which obviously implies $f \in \mathcal{Q}_p$ and so, $\mathcal{B} \subseteq \mathcal{Q}_p$. On the other hand, if $f \in \mathcal{Q}_p$ then (1.3) infers $f \in \mathcal{B}$, and consequently, $\mathcal{Q}_p \subseteq \mathcal{B}$. Thus (i) is true.

Secondly, the gap series $f_1(z) = \sum_{k=0}^{\infty} z^{2^k}$ shows (ii) via Theorem 1.2.1.

Thirdly, if $f_2(z) = \sum_{k=0}^{\infty} 2^{k/2} z^{2^k}$, then Theorem 1.2.1 is applied to show that $f_2 \in \mathcal{Q}_p \setminus \mathcal{Q}_0$ for any $p \in (0,1]$. Furthermore, if $0 < p < q \le 1$, then $f_3(z) = \sum_{k=0}^{\infty} 2^{k(p-1)/2} z^{2^k}$ belongs to $\mathcal{Q}_q \setminus \mathcal{Q}_p$, by Theorem 1.2.1. Therefore, (iii) is proved.

1.3 Image Area

It is clear that every $\mathcal{Q}_0$-function has a finite image area (counting multiplicity). In order to find out a corresponding $\mathcal{Q}_p$-version, we introduce a square root of weighted image area integral of $f \in \mathcal{H}$ as follows:

$$G_p(f,w) = \left(p \int_0^1 \left(\int_{\mathbf{D}(w,r)} |f'(z)|^2 dm(z) \right) (1-r)^{p-1} dr \right)^{1/2}$$

for $w \in \mathbf{D}$ and $p > 0$. Moreover, let $G_0(f,w) = \lim_{p \searrow 0} G_p(f,w)$.

Theorem 1.3.1. *Let $p \in [0,\infty)$ and $f \in \mathcal{H}$. Then $f \in \mathcal{Q}_p$ if and only if $\sup_{w\in\mathbf{D}} G_p(f,w) < \infty$.*

Proof. Note that for $z, w \in \mathbf{D}$, and $p \in (0,\infty)$,

$$(1 - |\sigma_w(z)|^2)^p = p \int_{|\sigma_w(z)|^2}^1 (1-r)^{p-1} dr.$$

An application of Fubini's theorem to $F_p(f,w)$ implies the desired result.

For $f \in \mathcal{H}$ and $a \in \mathbf{C}$, let $n(f,a)$ be the number of the a-points z_k in $\mathbf{D}$ such that $f(z_k) = a$. Whenever $n(f,a) \leq 1$, f is called univalent function.

Corollary 1.3.1. *Let $p \in (0,\infty)$ and $f \in \mathcal{H}$ satisfy*

$$N(f) = \sup_{b\in\mathbf{C}} \int_{|a-b|\leq 1} n(f,a)dm(a) < \infty. \tag{1.6}$$

Then $f \in \mathcal{Q}_p$ if and only if $f \in \mathcal{B}$.

Proof. Suppose now that $f \in \mathcal{H}$ obeys (1.6). Due to $\mathcal{Q}_p \subseteq \mathcal{B}$, the proof will be finished if we can verify that $f \in \mathcal{B}$ implies $f \in \mathcal{Q}_p$. When f lies in $\mathcal{B}$, we always have

$$|||f|||_{\mathcal{B}} = |f(0)| + \|f\|_{\mathcal{B}} < \infty,$$

and so

$$|f(z)| \leq \frac{|||f|||_{\mathcal{B}}}{2} \log \frac{2}{1-|z|}, \quad z \in \mathbf{D}. \tag{1.7}$$

For $w \in \mathbf{D}$, set $f_w = f \circ \sigma_w - f(w)$. Since $\mathcal{B}$ is conformally invariant, (1.7) implies

$$\sup_{|z|=r} |f_w(z)| \leq \frac{\|f\|_{\mathcal{B}}}{2} \log \frac{2}{1-r}, \quad r \in (0,1). \tag{1.8}$$

If f satisfies (1.6), then so does f_w with $N(f_w) = N(f)$. Notice that for $R \in (0,\infty)$, the disk $\{z \in \mathbf{C} : |z| \leq R\}$ may be covered by $4[1+R]$ (where $[1+R]$ means the integer part of $1+R$) of the disks: $\{z \in \mathbf{C} : |z - z_j| \leq 1\}$. Accordingly,

$$\int_{\mathbf{D}(0,r)} |f_w'|^2 dm \le \int_{|a|\le \sup_{|z|=r}|f_w(z)|} n(f_w,a)dm(a)$$
$$\le 4N(f)\Big(1+\sup_{|z|=r}|f_w(z)|\Big)^2.$$

This, together with (1.8), gives

$$(G_p(f,w))^2 \preceq N(f)\|f\|_{\mathcal{B}}^2 \int_0^1 \Big(1+\log\frac{2}{1-r}\Big)^2 (1-r)^{p-1}dr.$$

Thus, $f \in \mathcal{Q}_p$ is derived from Theorem 1.3.1.

1.4 Higher Derivatives

We need a lemma involving a useful integral estimation.

Lemma 1.4.1. *Let $c, t+1 \in (0,\infty)$. Then*

$$\int_{\mathbf{D}} \frac{(1-|v|^2)^t}{|1-\bar{v}z|^{t+2+c}} dm(v) \approx (1-|z|^2)^{-c}, \quad z \in \mathbf{D}.$$

Proof. For $z, v \in \mathbf{D}$ and $\alpha = (2+t+c)/2 > 0$, we have

$$\frac{1}{(1-\bar{v}z)^\alpha} = \sum_{n=0}^{\infty} \frac{\Gamma(n+\alpha)}{n!\Gamma(\alpha)} \bar{v}^n z^n.$$

Here and henceforward, $\Gamma(\cdot)$ denotes the classical gamma function. By Stirling's formula, we find

$$\frac{(\Gamma(n+\alpha))^2}{n!\Gamma(n+t+2)} \approx n^{c-1},$$

thus

$$\int_{\mathbf{D}} \frac{(1-|v|^2)^t}{(1-z\bar{v})^{2\alpha}} dm(v) = \frac{\Gamma(t+1)}{(\Gamma(\alpha))^2} \sum_{n=0}^{\infty} \frac{(\Gamma(n+\alpha))^2}{n!\Gamma(n+t+2)} |z|^{2n} \approx (1-|z|^2)^{-c}.$$

With this lemma, we can establish a higher derivative criterion for $\mathcal{Q}_p$.

Theorem 1.4.1. *Let $n \in \mathbf{N}$, $p \in [0,\infty)$ and $f \in \mathcal{H}$. Then $f \in \mathcal{Q}_p$ if and only if*

$$H_p(f,n) = \Big(\sup_{w\in\mathbf{D}} \int_{\mathbf{D}} |f^{(n)}(z)|^2 (1-|\sigma_w(z)|^2)^p (1-|z|^2)^{2n-2} dm(z)\Big)^{1/2} < \infty.$$

Proof. We break the proof into four steps.

Step 1: the case $p=0$. The result will follow directly from

$$|f^{(n)}(0)|^2 + (F_{\alpha+2}(f^{(n)},0))^2 \approx \int_{\mathbf{D}} |f^{(n)}(z)|^2 (1-|z|)^\alpha dm(z), \qquad (1.9)$$

where $\alpha \in [0,\infty)$, $f \in \mathcal{H}$, and $n \in \mathbf{N} \cup \{0\}$.

To prove (1.9), let $f^{(n)}(z) = \sum_{k=0}^{\infty} a_k z^k$. Then $f^{(n+1)}(z) = \sum_{k=1}^{\infty} k a_k z^{k-1}$. Using Parseval's formula we get

$$(F_{\alpha+2}(f^{(n)},0))^2 \approx \sum_{k=1}^{\infty} k^2 |a_k|^2 B(2k, \alpha+3)$$

and

$$\int_{\mathbf{D}} \frac{|f^{(n)}(z)|^2}{(1-|z|)^{-\alpha}} dm(z) \approx |f^{(n)}(0)|^2 B(2,\alpha+1) + \sum_{k=1}^{\infty} |a_k|^2 B(2k+2, \alpha+1).$$

Here and afterwards, $B(\cdot,\cdot)$ denotes the classical Beta function. Since Stirling's formula gives

$$B(2k+2,\alpha+1) \approx k^{-(1+\alpha)}; \quad B(2k,\alpha+3) \approx k^{-(\alpha+3)},$$

the desired comparability (1.9) follows.

Step 2: we prove that as for $f \in \mathcal{H}$, $f \in \mathcal{B}$ if and only if

$$M(f,n) = \sup_{z\in\mathbf{D}} (1-|z|^2)^n |f^{(n)}(z)| < \infty$$

for the integer $n \geq 2$.

Let $f \in \mathcal{B}$ and $f(0) = 0$. The Taylor expansion of f at 0 shows

$$f(z) = \frac{1}{\pi} \int_{\mathbf{D}} \frac{(1-|u|^2) f'(u)}{\bar{u}(1-z\bar{u})^2} dm(u). \tag{1.10}$$

Derivating the formula (1.10) gives

$$(1-|z|^2)^n f^{(n)}(z) = \frac{(n+1)!(1-|z|^2)^n}{\pi} \int_{\mathbf{D}} \frac{\bar{u}^{n-1}(1-|u|^2)f'(u)}{(1-z\bar{u})^{n+2}} dm(u).$$

This, together with Lemma 1.4.1, implies $M(f,n) \preceq \|f\|_{\mathcal{B}} < \infty$.

Conversely, suppose that f satisfies $M(f,n) < \infty$ and $f(0) = f'(0) = \cdots = f^{(n-1)}(0) = 0$. From the proof of Lemma 1.4.1 it follows that

$$f(0) = \frac{(n+1)}{\pi} \int_{\mathbf{D}} f(u)(1-|u|^2)^n dm(u),$$

which implies (replacing f with $f \circ \sigma_z$, $z \in \mathbf{D}$)

$$f(z) = \frac{(n+1)}{\pi} \int_{\mathbf{D}} \frac{f(v)(1-|v|^2)^n}{(1-z\bar{v})^{n+2}} dm(v).$$

Moreover, changing f into $f^{(n)}$ in the above formula infers

$$f^{(n)}(z) = \frac{(n+1)}{\pi} \int_{\mathbf{D}} \frac{f^{(n)}(v)(1-|v|^2)^n}{(1-z\bar{v})^{n+2}} dm(v).$$

By taking the line integral from 0 to z and noticing $f^{(n-1)}(0)=0$, we get

$$f^{(n-1)}(z)=\frac{1}{\pi}\int_{\mathbf{D}}\frac{(1-(1-z\bar{v})^{n+1})f^{(n)}(v)(1-|v|^2)^n}{\bar{v}(1-z\bar{v})^{n+2}}dm(v).$$

This, together with a simple calculation, yields

$$(1-|z|^2)^{n-1}|f^{(n-1)}(z)|\preceq\sup_{v\in\mathbf{D}}(1-|v|^2)^n|f^{(n)}(v)|.$$

Continuing this process, we can finally obtain $\|f\|_{\mathcal{B}}\preceq M(f,n)$.

Step 3: the case $p\in(1,\infty)$. At this point, $\mathcal{Q}_p=\mathcal{B}$. If $f\in\mathcal{Q}_p$, then $f\in\mathcal{B}$ and hence $M(f,n)<\infty$. Lemma 1.4.1 is applied to produce $H_p(f,n)\preceq M(f,n)$. Conversely, if $H_p(f,n)<\infty$, then by the subharmonic property of $|f^{(n)}|^2$, we easily find out that

$$(M(f,n))^2\preceq\sup_{w\in\mathbf{D}}\int_{\mathbf{D}(w,1/2)}\frac{|f^{(n)}(z)|^2(1-|\sigma_w(z)|^2)^2}{(1-|z|^2)^{2-2n}}dm(z). \tag{1.11}$$

Thus, the above second step shows $f\in\mathcal{B}$, i.e., $f\in\mathcal{Q}_p$.

Step 4: the case $p\in(0,1]$. Observe that if $n=1$ then the result is true, owing to Theorem 1.1.1. Suppose now that the direction that $f\in\mathcal{Q}_p$ implies $H_p(f,n)<\infty$ holds for some fixed integer $n\geq 2$. We must prove that $H_p(f,n+1)<\infty$ whenever $f\in\mathcal{Q}_p$. Now assume that $f\in\mathcal{Q}_p$. Note that (1.9) indicates that for $g\in\mathcal{H}$,

$$(F_{\alpha+2}(g,0))^2+|g'(0)|^2\approx\int_{\mathbf{D}}|g(z)|^2(1-|z|^2)^{\alpha}dm(z) \tag{1.12}$$

for $0<\alpha<\infty$. Taking $g(z)=f^{(n)}(z)/(1-\bar{w}z)^p$, $w\in\mathbf{D}$ and $\alpha=2n-2+p$ in (1.12), putting

$$R_1(f,w,z)=\frac{|f^{(n+1)}(z)|^2}{|1-\bar{w}z|^{2p}};$$

$$R_2(f,w,z)=\frac{p^2|w|^2|f^{(n)}(z)|^2}{|1-\bar{w}z|^{2p+2}};$$

$$R_3(f,w,z)=2Re\left(\frac{pwf^{(n+1)}(z)\overline{f^{(n)}(z)}}{(1-\bar{w}z)^p(1-w\bar{z})^{p+1}}\right),$$

and letting

$$T_j(f,w)=\int_{\mathbf{D}}R_j(f,w,z)(1-|z|^2)^{2n+p}(1-|w|^2)^p dm(z)$$

for $j=1,2,3$, we have

$$\begin{aligned}\sum_{j=1}^{3} T_j(f,w) &= \int_{\mathbf{D}} \sum_{j=1}^{3} R_j(f,w,z)(1-|z|^2)^{2n+p}(1-|w|^2)^p dm(z)\\ &\preceq \int_{\mathbf{D}} \frac{|f^{(n)}(z)|^2(1-|w|^2)^p(1-|z|^2)^{p+2n-2}}{|1-\bar{w}z|^{2p}} dm(z)\\ &= \int_{\mathbf{D}} |f^{(n)}(z)|^2(1-|\sigma_w(z)|^2)^p(1-|z|^2)^{2n-2} dm(z)\\ &\le (H_p(f,n))^2.\end{aligned}$$

Since $f \in \mathcal{B}$, it follows from Step 2 and Lemmas 1.4.1 that

$$\begin{aligned}|T_3(f,w)| &\le 2\int_{\mathbf{D}} \frac{|f^{(n)}(z)f^{(n+1)}(z)|(1-|z|^2)^{2n+p}(1-|a|^2)^p}{|1-\bar{w}z|^{2p+1}} dm(z)\\ &\preceq M(f,n)M(f,n+1)\end{aligned}$$

and

$$\begin{aligned}T_2(f,w) &\preceq \int_{\mathbf{D}} \frac{|f^{(n)}(z)|^2(1-|z|^2)^{p+2n-2}(1-|w|^2)^p}{|1-\bar{w}z|^{2p}} dm(z)\\ &= \int_{\mathbf{D}} |f^{(n)}(z)|^2(1-|\sigma_w(z)|^2)^p(1-|z|^2)^{2n-2} dm(z)\\ &\le (H_p(f,n))^2.\end{aligned}$$

Consequently,

$$\begin{aligned}T_1(f,w) &\preceq (H_p(f,n))^2 + T_2(f,w) + |T_3(f,w)|\\ &\preceq (H_p(f,n))^2 + M(f,n)M(f,n+1).\end{aligned}$$

If $f \in \mathcal{Q}_p$ then $f \in \mathcal{B}$ and $H_p(f,n) < \infty$ (by the inductive assumption), and hence $\sup_{w\in\mathbf{D}} T_1(f,w) < \infty$ (by the estimate on $T_1(f,w)$ obtained as above). Since $\sup_{w\in\mathbf{D}} T_1(f,w) = (H_p(f,n+1))^2$, we reach $H_p(f,n+1) < \infty$. Therefore, by the induction of n, that $f \in \mathcal{Q}_p$ implies $H_p(f,n) < \infty$ is valid for all $n > 1$.

The opposite direction that $H_p(f,n) < \infty$ implies $f \in \mathcal{Q}_p$ can be verified similarly. In fact, when $n = 1$, the result follows from Theorem 1.1.1. Suppose that $H_p(f,n) < \infty$ implies $f \in \mathcal{Q}_p$ holds for some fixed integer $n \ge 2$. If $H_p(f,n+1) < \infty$, then from (1.11) it follows that $M(f,n+1) < \infty$ and so $M(f,n) < \infty$ by the second part of Step 2. In (1.12) we make the substitution: $g(z) = f^{(n)}(z)/(1-\bar{w}z)^p$, $w \in \mathbf{D}$, $\alpha = 2n-2+p$, and use the equivalence in Step 2 to deduce

$$\begin{aligned}(H_p(f,n))^2 &\preceq (M(f,n))^2 + (M(f,n+1))^2\\ &\quad + \sup_{w\in\mathbf{D}} \int_{\mathbf{D}} \left|\frac{d}{dz}\left(\frac{f^{(n)}(z)}{(1-\bar{w}z)^p}\right)\right|^2 \frac{(1-|z|^2)^{2n+p}}{(1-|w|^2)^{-p}} dm(z)\\ &\preceq (M(f,n))^2 + (M(f,n+1))^2 + \sup_{w\in\mathbf{D}} \sum_{j=5}^{7} T_j(f,w),\end{aligned}$$

where

$$
\begin{aligned}
T_5(f,w) &= \int_{\mathbf{D}} \frac{|f^{(n+1)}(z)|^2(1-|z|^2)^{2n+p}(1-|w|^2)^p}{|1-\bar{w}z|^{2p}}\,dm(z);\\
T_6(f,w) &= \int_{\mathbf{D}} \frac{|f^{(n)}(z)|^2(1-|z|^2)^{2n+p}(1-|w|^2)^p}{|1-\bar{w}z|^{2p+2}}\,dm(z);\\
T_7(f,w) &= \int_{\mathbf{D}} \frac{|f^{(n+1)}(z)f^{(n)}(z)|(1-|z|^2)^{2n+p}(1-|w|^2)^p}{|1-\bar{w}z|^{2p+1}}\,dm(z).
\end{aligned}
$$

However, Lemma 1.4.1, together with $f \in \mathcal{B}$, gives

$$
\begin{aligned}
T_5(f,w) &\preceq (H_p(f,n+1))^2;\\
T_6(f,w) &\preceq (M(f,n))^2;\\
T_7(f,w) &\preceq M(f,n+1)M(f,n).
\end{aligned}
$$

As a result, $H_p(f,n) < \infty$ and hence $f \in \mathcal{Q}_p$ (due to the inductive assumption). Therefore, the proof is complete.

Notes

1.1 Theorem 1.1.1 was first proved by Aulaskari, Xiao and Zhao [27] in the case $p \in (0,1]$, and by Stroethoff [121] and Xiao [133] for the Bloch space. But, the proof idea of Theorem 1.1.1 originates from Aulaskari-Stegenga-Xiao [24]. Of course, the concept BMO is due to John-Nirenberg [82]. As for Corollary 1.1.1, see also Baernstein [30], Axler [29], Aulaskari-Stegenga-Xiao [24] and Aulaskari-Lappan-Xiao-Zhao [21]. For the definitions and some basic properties of the $\mathcal{Q}_p$ spaces in several complex variables, see also Andersson-Carlsson [9], Ouyang-Yang-Zhao [98], Latvala [85], Hu-Shi-Zhang [79]. Meantime, the interested reader is referred to Gürlebeck-Kähler-Shapiro-Tovar [73], Cnops-Delanghe [40] and Cnops-Delanghe-Gürlebeck-Shapiro [41] for an account of $\mathcal{Q}_p$-spaces in Clifford analysis. Concerning a survey or a mini-course or a research report on $\mathcal{Q}_p$, one mentions Essén-Xiao [63] or Essén [59] or Aulaskari [13].

1.2 Theorem 1.2.1 is in Aulaskari-Xiao-Zhao [27] and Pommerenke [100], respectively. However, its proof presented in the text is quite direct and related to the one in Mateljevic-Pavlovic [90]. As is done in Corollary 1.2.1, the criterion stated in Theorem 1.2.1 provides a method to construct the highly non-univalent $\mathcal{Q}_p$-functions. Moreover, given f as in Theorem 1.2.1. If $f \in \mathcal{B}$ then by (ii), $\sup_k |a_k| < \infty$ holds, and hence for $p > 1$, noticing that the number of a_j is less than $1 + \log_c 2$ when $2^k \le n_j < 2^{k+1}$, one has (1.4). However, condition (1.4) cannot ensure that $f \in \mathcal{B}$. In fact, choosing $a_j = j$ and $n_j = 2^j$ leads to that the induced function is not in $\mathcal{B}$, but satisfies (1.4).

In addition, the result that the spaces $\mathcal{B}$ and $\mathcal{Q}_2$ are the same (with the comparable norms) was first found by Xiao [133], and secondly extended by

Aulaskari and Lappan [20] to all $p > 1$. These facts have actually proved that $\mathcal{B}$ is the maximal space among all $\mathcal{Q}_p$ spaces. For a similar result, see also Rubel-Timoney [106] and Wulan-Wu [132]. However, the symbol $\mathcal{Q}_p$, which represents a new space, first appeared in Aulaskari-Xiao-Zhao [27]. Arazy, Fisher and Peetre defined Möbius invariant spaces of holomorphic functions on $\mathbf{D}$, and studied general properties of those spaces, in particular, worked out many characterizations of the minimal Möbius invariant space; see [12] and [10].

1.3 Theorem 1.3.1 characterizes $\mathcal{Q}_p$-functions in terms of the image area (with multiplicity). Unfortunately, the image area $\int_{\mathbf{D}(w,r)} |f'|^2 dm$ cannot be replaced by the real area $\int_{f(\mathbf{D}(w,r))} dm$ in general (constructing counterexample via Theorem 1.2.1 (i)). In the paper [68] by Gauthier and Xiao it is proved that $f \in \mathcal{B}$ if and only if

$$\sup_{w\in\mathbf{D}} \left(p \int_0^1 \left(\int_{f(\mathbf{D}(w,r))} dm \right) (1-r)^{p-1} dr \right)^{1/2} < \infty$$

for all $p > 0$.

It is worthwhile to point out that Corollary 1.3.1 generalizes Pommerenke's [101, Satz 1], but also reveals that the univalent functions cannot keep one $\mathcal{Q}_p$ class separate from another. For a further discussion, see also Aulaskari-Lappan-Xiao-Zhao [21], Wirths-Xiao [126] and Gauthier-Xiao [68].

1.4 Theorem 1.4.1 is the main result of [22] by Aulaskari-Nowak-Zhao. The $\mathcal{B}$-setting of Theorem 1.4.1 is known, and the proof presented in the second step, is essentially taken from Zhu [144]. However, in case of BMOA, Theorem 1.4.1 solves a question of Stroethoff posed in [121].

2. Composite Embedding

This chapter is devoted to study the smooth properties of the $\mathcal{Q}_p$-functions through characterizing a self-map ϕ of $\mathbf{D}$ such that the composition operator C_ϕ maps $\mathcal{Q}_p$ into $\mathcal{B}^\alpha$ (the Bloch-type space) and vice versa.

2.1 Existence of BiBloch-type Mappings

For $\alpha \in (0, \infty)$, let $\mathcal{B}^\alpha$ denote the Bloch-type space of all functions $f \in \mathcal{H}$ whose α-expansions have finite upper bound:

$$\|f\|_{\mathcal{B}^\alpha} = \sup_{z \in \mathbf{D}} (1 - |z|^2)^\alpha |f'(z)| < \infty.$$

Of course, the following result is well known: if $\alpha = 1$ or $\alpha \in (0, 1)$ then $\mathcal{B}^\alpha$ coincides with the Bloch space or the classical $(1 - \alpha)$-Lipschitz class (cf. [50, Theorem 5.1]). It is clear that $\mathcal{B}^\alpha$ is a Banach space with respect to the norm $|f(0)| + \|f\|_{\mathcal{B}^\alpha}$.

It is impossible to find such a holomorphic map $f : \mathbf{D} \to \mathbf{C}$ whose α-expansion has both positive upper and lower bounds: $(1 - |z|^2)^\alpha |f'(z)| \preceq 1$; $(1 - |z|^2)^\alpha |f'(z)| \succeq 1$ for all $z \in \mathbf{D}$. In fact, if, otherwise, there is such a map f, then one would have that $1/|f'(z)| \approx (1 - |z|^2)^\alpha$ as $|z| \to 1$ and consequently, $1/f' \equiv 0$. This simple but important observation leads to the following consideration.

Lemma 2.1.1. *Let $\alpha \in (0, \infty)$ and let $f(z) = \sum_{j=1}^{\infty} a_j z^{n_j}$ belong to HG. Then $f \in \mathcal{B}^\alpha$ if and only if $\sup_{j \in \mathbf{N}} |a_j| n_j^{1-\alpha} < \infty$.*

Proof. This is due to Yamashita [140]. However, we include a proof for completeness. Consulting Theorem 1.2.1 (ii), we find it enough to show the 'if' part. Suppose $\sup_{k \in \mathbf{N}} |a_k| n_k^{1-\alpha} < \infty$, and $K = K(n) = \max\{k : n_k \le n\}$. Thus

$$\frac{1}{n^\alpha} \sum_{n_j \le n} n_j^\alpha = \left(\frac{n_K}{n}\right)^\alpha \sum_{j=0}^{K-1} \left(\frac{n_{K-j}}{n_K}\right)^\alpha \preceq 1.$$

Hence

$$\frac{|zf'(z)|}{1-|z|} \preceq \Big(\sum_{k=0}^{\infty}|z|^k\Big)\Big(\sum_{j=0}^{\infty}n_j^{\alpha}|z|^{n_j}\Big)$$
$$\preceq \sum_{n=1}^{\infty}\Big(\sum_{n_j\le n}n_j^{\alpha}\Big)|z|^n \preceq \frac{|z|}{(1-|z|)^{1+\alpha}},$$

which gives $f \in \mathcal{B}^{\alpha}$.

Theorem 2.1.1. *Let $\alpha \in (0,\infty)$. Then there are two holomorphic maps $f_1, f_2 : \mathbf{D} \to \mathbf{C}$ such that*

$$|f_1'(z)| + |f_2'(z)| \approx (1-|z|^2)^{-\alpha}, \quad z \in \mathbf{D}. \tag{2.1}$$

Proof. For a large number $q \in \mathbf{N}$, choose a gap series:

$$f_{\alpha}(z) = \sum_{j=0}^{\infty} q^{j(\alpha-1)} z^{q^j}, \quad z \in \mathbf{D}.$$

Then, apply Lemma 2.1.1 with $a_j = q^{j(\alpha-1)}$ and $n_j = q^j$ to infer that $(1-|z|^2)|f_{\alpha}'(z)| \preceq 1$ holds for all $z \in \mathbf{D}$. Furthermore, let us verify

$$(1-|z|^2)^{\alpha}|f_{\alpha}'(z)| \succeq 1, \quad 1-q^{-k} \le |z| \le 1-q^{-(k+1/2)}, \quad k \in \mathbf{N}. \tag{2.2}$$

Observe that for any $z \in \mathbf{D}$,

$$|f_{\alpha}'(z)| \ge q^{k\alpha}|z|^{q^k} - \sum_{j=0}^{k-1} q^{j\alpha}|z|^{q^j} - \sum_{j=k+1}^{\infty} q^{j\alpha}|z|^{q^j} = T_1 - T_2 - T_3.$$

And then, fix a z with $|z| \in [1-q^{-k}, 1-q^{-(k+1/2)}]$, $k \in \mathbf{N}$, and put $x = |z|^{q^k}$. Thus

$$(1-q^{-k})^{q^k} \le x \le \big((1-q^{-(k+1/2)})^{q^{k+1/2}}\big)^{q^{-1/2}}.$$

If q is large enough, then for $k \ge 1$ one has

$$\frac{1}{3} \le x \le \Big(\frac{1}{2}\Big)^{q^{-1/2}}, \tag{2.3}$$

and hence $T_1 \ge q^{k\alpha}/3$. Since it is easy to establish

$$T_2 \le \sum_{j=0}^{k-1} q^{j\alpha} \le \frac{q^{k\alpha}}{q^{\alpha}-1},$$

it remains to deal with the third term T_3. Noting that

$$|z|^{q^n(q-1)} \le |z|^{q^{k+1}(q-1)}, \quad n \ge k+1,$$

namely, in T_3 the quotient of two successive terms is not greater than the ratio of the first two terms, one finds that the series of T_3 is controlled by the geometric series having the same first two terms. Accordingly (2.3) is applied to produce

$$T_3 \le q^{(k+1)\alpha}|z|^{q^{k+1}} \sum_{j=0}^{\infty} \left(q^\alpha |z|^{(q^{k+2}-q^{k+1})}\right)^j = \frac{q^{(k+1)\alpha}|z|^{q^{k+1}}}{1 - q^\alpha |z|^{(q^{k+2}-q^{k+1})}}$$

$$= \frac{q^{k\alpha}q^\alpha x^q}{1 - q^\alpha x^{(q^2-q)}} \le \frac{q^{k\alpha}q^\alpha 2^{-q^{1/2}}}{1 - q^\alpha 2^{-(q^{3/2}-q^{1/2})}}.$$

The preceding estimates for T_1, T_2 and T_3 imply

$$|f'_\alpha(z)| \ge \frac{q^{k\alpha}}{4} = \frac{(q^\alpha)^{k+1/2}}{4q^{\alpha/2}} \ge \frac{1}{4q^{\alpha/2}(1-|z|)^\alpha},$$

reaching (2.2).

In a completely similar manner one can prove that if q is a large natural number, for example $q = m^2$ where m is a large natural number, and if

$$g_\alpha(z) = \sum_{j=0}^{\infty} q^{(j+1/2)(\alpha-1)} z^{q^j}, \quad z \in \mathbf{D},$$

then $(1-|z|^2)^\alpha |g'_\alpha(z)| \preceq 1$ for all $z \in \mathbf{D}$ (owing to Lemma 2.1.1) and

$$(1-|z|^2)^\alpha |g'_\alpha(z)| \succeq 1, \quad 1 - q^{-(k+1/2)} \le |z| \le 1 - q^{-(k+1)}, \quad k \in \mathbf{N}. \tag{2.4}$$

Of course, (2.2) and (2.4) yield (2.1) unless it occurs that f'_α and g'_α share a zero in $\{z \in \mathbf{D} : |z| < 1 - q^{-1}\}$, in which case one can replace g_α with $g_\alpha(\zeta z)$ for an appropriate $\zeta \in \mathbf{T}$ (thanks to $f'_\alpha(0) = 1$). Therefore we are done.

Corollary 2.1.1. *Let $\alpha \in (0, \infty)$. Then $\mathcal{B}^\alpha$ is conformally invariant if and only if $\alpha = 1$.*

Proof. Sufficiency is clear. Concerning necessity, one proceeds below. Let $\mathcal{B}^\alpha$ be conformally invariant. Then $\|f \circ \sigma_w\|_{\mathcal{B}^\alpha} = \|f\|_{\mathcal{B}^\alpha}$ for all $f \in \mathcal{B}^\alpha$ and $w \in \mathbf{D}$. By Theorem 2.1.1, there are $f_1, f_2 \in \mathcal{B}^\alpha$ such that $(1-|z|^2)^\alpha(|f'_1(z)| + |f'_2(z)|) \approx 1$ for all $z \in \mathbf{D}$. Consequently, for $z, w \in \mathbf{D}$ one has

$$1 \approx (1-|\sigma_w(z)|^2)^\alpha \Big(|f'_1(\sigma_w(z))| + |f'_2(\sigma_w(z))|\Big)$$

$$= \left(\frac{1-|w|^2}{|1-\bar{w}z|^2}\right)^{\alpha-1} (1-|z|^2)^\alpha \Big(|(f_1 \circ \sigma_w)'(z)| + |(f_2 \circ \sigma_w)'(z)|\Big).$$

This, together with the invariance hypothesis on $\mathcal{B}^\alpha$, implies $1 \approx (1-|w|^2)^{\alpha-1}$ for all $w \in \mathbf{D}$, forcing $\alpha = 1$.

2.2 Boundedness and Compactness

Any holomorphic map $\phi : \mathbf{D} \to \mathbf{D}$ gives rise to a composition operator $C_\phi : \mathcal{H} \to \mathcal{H}$ defined by $C_\phi f = f \circ \phi$, the composition operator induced by ϕ. One of the central problems on composition operators is to know when C_ϕ maps between two subclasses of $\mathcal{H}$ and in fact to relate function theoretic properties of ϕ to operator theoretic properties (e.g. boundedness and compactness) of C_ϕ sending one subclass to another.

For an $r \in (0,1)$ and a holomorphic self-map ϕ of $\mathbf{D}$, put $\Omega_r = \{z \in \mathbf{D} : |\phi(z)| > r\}$, and $\Delta_r = \{z \in \mathbf{D} : |z| > r\}$. The characteristic function of a set $\mathbf{E} \subseteq \mathbf{D}$ is denoted by $1_\mathbf{E}$. Also, $\mathbf{B}_X$ stands for the unit ball of the given Banach space X. By H^∞ one means the set of all functions $f \in \mathcal{H}$ with $\|f\|_{H^\infty} = \sup_{z\in\mathbf{D}} |f(z)| < \infty$.

Theorem 2.2.1. *Let $\alpha, p \in (0,\infty)$ and let $\phi : \mathbf{D} \to \mathbf{D}$ be holomorphic. Then*

(i) $C_\phi : \mathcal{B}^\alpha \to \mathcal{Q}_p$ *exists as a bounded operator if and only if*

$$\sup_{w\in\mathbf{D}} \int_\mathbf{D} \frac{|\phi'(z)|^2}{(1-|\phi(z)|^2)^{2\alpha}}(1-|\sigma_w(z)|^2)^p dm(z) < \infty. \tag{2.5}$$

(ii) $C_\phi : \mathcal{B}^\alpha \to \mathcal{Q}_p$ *exists as a compact operator if and only if $\phi \in \mathcal{Q}_p$ and*

$$\lim_{r\to 1} \sup_{w\in\mathbf{D}} \int_\mathbf{D} \frac{|\phi'(z)|^2}{(1-|\phi(z)|^2)^{2\alpha}}(1-|\sigma_w(z)|^2)^p 1_{\Omega_r}(z) dm(z) = 0. \tag{2.6}$$

(iii) $C_\phi : \mathcal{Q}_p \to \mathcal{B}^\alpha$ *exists as a bounded operator if and only if*

$$\sup_{z\in\mathbf{D}} \frac{|\phi'(z)|}{1-|\phi(z)|^2}(1-|z|^2)^\alpha < \infty. \tag{2.7}$$

(iv) $C_\phi : \mathcal{Q}_p \to \mathcal{B}^\alpha$ *exists as a compact operator if and only if $\phi \in \mathcal{B}^\alpha$ and*

$$\lim_{r\to 1} \sup_{z\in\mathbf{D}} \frac{|\phi'(z)|}{1-|\phi(z)|^2}(1-|z|^2)^\alpha 1_{\Omega_r}(z) = 0. \tag{2.8}$$

Proof. (i) This follows obviously from Theorems 2.1.1 and 1.1.1, as well as a simple calculation.

(ii) Let $\phi \in \mathcal{Q}_p$ and let (2.6) hold. By Theorem 1.1.1, we are required to show that if $\{f_n\} \subset \mathbf{B}_{\mathcal{B}^\alpha}$ converges to 0 uniformly on compact subsets of $\mathbf{D}$ then $\{\|C_\phi f_n\|_{\mathcal{Q}_p}\}$ converges to 0. For each $r \in (0,1)$ set $\tilde{\Omega}_r = \mathbf{D} \setminus \Omega_r$. So $\{f_n' \circ \phi\}$ tends to 0 uniformly on $\tilde{\Omega}_r$. And hence by Theorem 1.1.1, for every $\epsilon > 0$ there is an integer $N > 1$ such that as $n \geq N$,

$$\sup_{w\in\mathbf{D}} \int_\mathbf{D} |(C_\phi f_n)'(z)|^2 (1-|\sigma_w(z)|^2)^p 1_{\tilde{\Omega}_r}(z) dm(z) \preceq \epsilon \|\phi\|^2_{\mathcal{Q}_p}.$$

Meanwhile, from (2.6) and the growth of the derivatives of $\mathcal{B}^\alpha$-functions one derives that for every $\epsilon > 0$ there exists a $\delta \in (0,1)$ such that as $r \in [\delta, 1)$,

$$\sup_{w\in\mathbf{D}}\int_{\mathbf{D}}|(C_\phi f_n)'(z)|^2(1-|\sigma_w(z)|^2)^p 1_{\Omega_r}(z)dm(z)<\epsilon.$$

These inequalities are combined with Theorem 1.1.1 to imply that $\|C_\phi f_n\|_{\mathcal{Q}_p}\to 0$ as $n\to\infty$.

Conversely, let $C_\phi:\mathcal{B}^\alpha\to\mathcal{Q}_p$ be compact. It is clear that $\phi\in\mathcal{Q}_p$. So, it must be shown that (2.6) holds. Take $f_n(z)=z^n$ resp. $f_n(z)=z^n/n^{1-\alpha}$ if $\alpha\in[1,\infty)$ resp. $\alpha\in(0,1)$. Without loss of generality, we only consider $\alpha\in(0,1)$. Since $\{f_n\}$ is norm bounded in $\mathcal{B}^\alpha$ and it converges to 0 uniformly on compact subsets of $\mathbf{D}$, $\|\phi^n\|_{\mathcal{Q}_p}\to 0$. Applying Theorem 1.1.1, we find that for every $\epsilon>0$, there is an integer $N>1$ such that as $n\geq N$,

$$n^{2\alpha}\sup_{w\in\mathbf{D}}\int_{\mathbf{D}}|\phi(z)|^{2n-2}|\phi'(z)|^2(1-|\sigma_w(z)|^2)^p dm(z)<\epsilon;$$

thus for each $r\in(0,1)$,

$$N^{2\alpha}r^{2N-2}\sup_{w\in\mathbf{D}}\int_{\mathbf{D}}|\phi'(z)|^2(1-|\sigma_w(z)|^2)^p 1_{\Omega_r}(z)dm(z)<\epsilon.$$

Through taking $r\geq N^{-\frac{\alpha}{N-1}}$, we get

$$\sup_{w\in\mathbf{D}}\int_{\mathbf{D}}|\phi'(z)|^2(1-|\sigma_w(z)|^2)^p 1_{\Omega_r}(z)dm(z)<\epsilon. \tag{2.9}$$

Keeping (2.9) in mind, we show that for every $f\in\mathbf{B}_{\mathcal{B}^\alpha}$ and for every $\epsilon>0$, there is a $\delta=\delta(f,\epsilon)$ (depending on f and ϵ) such that as $r\in[\delta,1)$,

$$T(f,\phi,p,r)=\sup_{w\in\mathbf{D}}\int_{\mathbf{D}}|(C_\phi f)'(z)|^2(1-|\sigma_w(z)|^2)^p 1_{\Omega_r}(z)dm(z)<\epsilon. \tag{2.10}$$

As a matter of fact, if $f_t(z)=f(tz)$ for $f\in\mathbf{B}_{\mathcal{B}^\alpha}$ and $t\in(0,1)$, then $f_t\to f$ uniformly on compact subsets of $\mathbf{D}$ as $t\to 1$. Since $C_\phi:\mathcal{B}^\alpha\to\mathcal{Q}_p$ is compact, $\|f_t\circ\phi-f\circ\phi\|_{\mathcal{Q}_p}\to 0$ as $t\to 1$. Furthermore, from Theorem 1.1.1 it yields that for every $\epsilon>0$ there is a $t\in(0,1)$ to insure

$$\sup_{w\in\mathbf{D}}\int_{\mathbf{D}}|(C_\phi f_t)'(z)-(C_\phi f)'(z)|^2(1-|\sigma_w(z)|^2)^p dm(z)<\epsilon.$$

Accordingly, by (2.9) one reaches

$$\begin{aligned}T(f,\phi,p,r)&\leq 2\epsilon+2\sup_{w\in\mathbf{D}}\int_{\mathbf{D}}|(C_\phi f_t)'(z)|^2(1-|\sigma_w(z)|^2)^p 1_{\Omega_r}(z)dm(z)\\&\leq 2\epsilon+2\|f_t'\|^2_{H^\infty}\sup_{w\in\mathbf{D}}\int_{\mathbf{D}}|\phi'(z)|^2(1-|\sigma_w(z)|^2)^p 1_{\Omega_r}(z)dm(z)\\&\leq 2\epsilon(1+\|f_t'\|^2_{H^\infty}).\end{aligned}$$

Since C_ϕ sends $\mathbf{B}_{\mathcal{B}^\alpha}$ to a relatively compact subset of $\mathcal{Q}_p$, there exists, for every $\epsilon > 0$, a finite collection of functions $f_1, f_2, .., f_N$ in $\mathbf{B}_{\mathcal{B}^\alpha}$ such that for each $f \in \mathbf{B}_{\mathcal{B}^\alpha}$ there is a $k \in \{1, 2, ..., N\}$ to guarantee

$$\sup_{w\in\mathbf{D}} \int_{\mathbf{D}} |(C_\phi f)'(z) - (C_\phi f_k)'(z)|^2 (1 - |\sigma_w(z)|^2)^p dm(z) < \epsilon.$$

Now (2.10) is used to induce a $\delta = \max_{1\le k\le N} \delta(f_k, \epsilon)$ so that as $r \in [\delta, 1)$,

$$\sup_{w\in\mathbf{D}} \int_{\mathbf{D}} |(C_\phi f_k)'(z)|^2 (1 - |\sigma_w(z)|^2)^p 1_{\Omega_r}(z) dm(z) < \epsilon;$$

thus

$$\sup_{f\in\mathbf{B}_{\mathcal{B}^\alpha}} \sup_{w\in\mathbf{D}} \int_{\mathbf{D}} |(C_\phi f)'(z)|^2 (1 - |\sigma_w(z)|^2)^p 1_{\Omega_r}(z) dm(z) \preceq \epsilon. \tag{2.11}$$

By Theorem 2.1.1 there are two functions $f_1, f_2 \in \mathbf{B}_{\mathcal{B}^\alpha}$ such that

$$|f_1'(\phi(z))|^2 + |f_2'(\phi(z))|^2 \succeq (1 - |\phi(z)|^2)^{-\alpha}$$

for all $z \in \mathbf{D}$. Thus (2.11) implies

$$\sup_{w\in\mathbf{D}} \int_{\mathbf{D}} \frac{|\phi'(z)|^2}{(1 - |\phi(z)|^2)^{2\alpha}} (1 - |\sigma_w(z)|^2)^p 1_{\Omega_r}(z) dm(z) \preceq \epsilon,$$

so that (2.6) follows.

(iii) Assume that $C_\phi : \mathcal{Q}_p \to \mathcal{B}^\alpha$ is bounded. Fix $z_0 \in \mathbf{D}$. Let $w = \phi(z_0)$ and pick $f_w(z) = -\log(1 - \bar{w}z)$. Then $f_w \in \mathcal{Q}_p$ with $\|f_w\|_{\mathcal{Q}_p} \preceq 1$, due to Corollary 1.3.1. By the boundedness of C_ϕ, we have

$$\|f_w\|_{\mathcal{Q}_p} \succeq |f_w(\phi(0)| + \|C_\phi f_w\|_{\mathcal{B}^\alpha} \succeq (1 - |z_0|^2)^\alpha |\phi'(z_0)| |1 - \bar{w}\phi(z_0)|^{-1},$$

which deduces (2.7).

Conversely, if (2.7) holds, then $\mathcal{Q}_p \subseteq \mathcal{B}$ gives that for any $f \in \mathcal{Q}_p$,

$$\|C_\phi f\|_{\mathcal{B}^\alpha} \preceq \|f\|_{\mathcal{Q}_p} \sup_{z\in\mathbf{D}} (1 - |z|^2)^\alpha \left(\frac{|\phi'(z)|}{1 - |\phi(z)|^2}\right) < \infty.$$

In other words, C_ϕ is a bounded operator from $\mathcal{Q}_p$ to $\mathcal{B}^\alpha$.

(iv) Let $C_\phi : \mathcal{Q}_p \to \mathcal{B}^\alpha$ be compact. Then $\phi \in \mathcal{B}^\alpha$ follows naturally. Now, suppose otherwise that the condition (2.8) fails. Then, there would be a number $\epsilon_0 > 0$ and a sequence $\{z_n\} \subset \mathbf{D}$ such that $(1 - |z_n|^2)^\alpha |\phi'(z_n)| (1 - |\phi(z_n)|^2)^{-1} \ge \epsilon_0$ whenever $|\phi(z_n)| > 1 - 1/n$. We may assume that $w_n = \phi(z_n)$ tend to a point $w_0 \in \mathbf{T}$. Put $f_n(z) = -\log(1 - \bar{w}_n z)$. Accordingly $\{f_n\}$ converges to f_0 uniformly on compact subsets of $\mathbf{D}$, where $f_0(z) = -\log(1 - \bar{w}_0 z)$. Appealing to these constructions, we obtain

$$\begin{aligned}\|C_\phi f_n - C_\phi f_0\|_{\mathcal{B}^\alpha} &\ge (1-|z_n|^2)^\alpha |(C_\phi f_n)'(z_n) - (C_\phi f_0)'(z_n)| \\ &= (1-|z_n|^2)^\alpha |\phi'(z_n)| \left| \frac{\bar{w}_n}{1-|w_n|^2} - \frac{\bar{w}_0}{1-\bar{w}_0 w_n)} \right| \\ &= \frac{(1-|z_n|^2)^\alpha |\phi'(z_n)|}{1-|\phi(z_n)|^2} \left| \frac{\bar{w}_n - \bar{w}_0}{1-\bar{w}_0 w_n} \right| \ge \epsilon_0\end{aligned}$$

for $n \in \mathbf{N}$, so $C_\phi f_n$ does not converge to $C_\phi f_0$ in norm. Hence $C_\phi : \mathcal{Q}_p \to \mathcal{B}^\alpha$ is not compact. This contradicts the above hypothesis.

On the other hand, let $\phi \in \mathcal{B}^\alpha$ and (2.8) hold. In order to show that $C_\phi : \mathcal{Q}_p \to \mathcal{B}^\alpha$ is compact, it suffices to verify that if $\{f_n\}$ is a bounded sequence in $\mathcal{Q}_p$ (i.e., $\sup_{n\in\mathbf{N}} \|f_n\|_{\mathcal{Q}_p} < \infty$) and if it converges to 0 uniformly on any compact subset of $\mathbf{D}$, then $\{\|C_\phi f_n\|_{\mathcal{B}^\alpha}\}$ approaches 0. By (2.8), we obtain that for any $\epsilon > 0$, there exists a $\delta \in (0,1)$ such that whenever $|\phi(z)| > \delta$,

$$(1-|z|^2)^\alpha |(C_\phi f_n)'(z)| \le \epsilon \sup_{n\in\mathbf{N}} \|f_n\|_{\mathcal{B}} \preceq \epsilon \sup_{n\in\mathbf{N}} \|f_n\|_{\mathcal{Q}_p}.$$

Yet, if $|\phi(z)| \le \delta$ then $(1-|\phi(z)|^2)^\alpha |f_n'(\phi(z))| \to 0$ owing to the fact that f_n tend to 0 uniformly on compact subsets of $\mathbf{D}$, and hence

$$(1-|z|^2)^\alpha |(C_\phi f_n)'(z)| \le \|\phi\|_{\mathcal{B}^\alpha} (1-\delta^2)^{-\alpha} (1-|\phi(z)|^2)^\alpha |f_n'(\phi(z))| \to 0.$$

The preceding estimates force $\|C_\phi f_n\|_{\mathcal{B}^\alpha} \to 0$. We are done.

It is clear that the condition $\|\phi\|_{H^\infty} < 1$ is a sufficient condition for Theorem 2.2.1 to be true. Moreover, the condition is also necessary for the boundedness of $C_\phi : \mathcal{Q}_p \to \mathcal{B}^\alpha$ whenever $\alpha \in (0,1)$ is assumed. This is because every $\mathcal{Q}_p$ contains an unbounded function $\log(1-z)$, but also because $\mathcal{B}^\alpha$ is a subspace of H^∞ for $\alpha \in (0,1)$.

Corollary 2.2.1. *Let $p, \alpha \in (0,\infty)$. Then*

(i) *$\mathcal{B}^\alpha$ is embedded boundedly into $\mathcal{Q}_p$ if and only if*

$$\sup_{w\in\mathbf{D}} \int_{\mathbf{D}} \frac{(1-|\sigma_w(z)|^2)^p}{(1-|z|^2)^{2\alpha}} dm(z) < \infty.$$

(ii) *$\mathcal{B}^\alpha$ is embedded compactly into $\mathcal{Q}_p$ if and only if*

$$\lim_{r\to 1} \sup_{w\in\mathbf{D}} \int_{\mathbf{D}} \frac{(1-|\sigma_w(z)|^2)^p}{(1-|z|^2)^{2\alpha}} 1_{\Omega_r}(z) dm(z) = 0.$$

(iii) *$\mathcal{Q}_p$ is embedded boundedly into $\mathcal{B}^\alpha$ if and only if $\alpha \ge 1$.*

(iv) *$\mathcal{Q}_p$ is embedded compactly into $\mathcal{B}^\alpha$ if and only if $\alpha > 1$.*

Proof. Theorem 2.2.1 with $\phi(z) = z$ will do the trick.

2.3 Geometric Characterizations

In this section we assume that the self-map ϕ of $\mathbf{D}$ is univalent, and provide geometric characterizations of when C_ϕ is either bounded or compact in this case. This requires some background on the hyperbolic metric.

Let $\rho_{\mathbf{D}}(\cdot,\cdot)$ denote the hyperbolic metric on $\mathbf{D}$, defined by

$$\rho_{\mathbf{D}}(z_1,z_2)=\inf\left\{\int_\gamma \frac{2|dz|}{1-|z|^2}:\ \gamma \textit{ is an arc in } \mathbf{D} \textit{ from } z_1 \textit{ to } z_2\right\}.$$

We note that a simple calculation shows $\rho_{\mathbf{D}}(0,z)=\log(1+|z|)/(1-|z|)$, and so

$$(1-|z|)^{-1}\le \exp(\rho_{\mathbf{D}}(0,z))\le 2(1-|z|)^{-1}. \tag{2.12}$$

This distance is invariant under $Aut(\mathbf{D})$, and therefore transfers to a natural conformally invariant metric on any simply connected proper subset G of $\mathbf{C}$. If $f:\mathbf{D}\to G$ is any conformal map, the hyperbolic distance on G is given by $\rho_G(w_1,w_2)=\rho_{\mathbf{D}}(z_1,z_2)$, where $w_j=f(z_j)$ for $j=1,2$. Furthermore, the $\rho_G(\cdot,\cdot)$ can also be computed by integrating a density function h_G over arcs in G. A change of variable argument shows that $h_G(f(z))=((1-|z|^2)|f'(z)|)^{-1}$. A useful geometric estimate for h_G that follows from the Koebe distortion theorem is that

$$\frac{1}{\delta_G(w)}\le h_G(w)\le \frac{2}{\delta_G(w)}, \tag{2.13}$$

where δ_G denotes the Euclidean distance of point $w\in G$ to $\mathbf{C}\setminus G$.

Theorem 2.3.1. *Let $\alpha,p\in(0,\infty)$ and let $\phi:\mathbf{D}\to\mathbf{D}$ be univalent. Then*

(i) *$C_\phi:\mathcal{B}^\alpha\to\mathcal{Q}_p$ exists as a bounded operator if and only if*

$$\sup_{w\in\phi(\mathbf{D})}\int_{\phi(\mathbf{D})}\frac{\exp\big(-p\rho_{\phi(\mathbf{D})}(w,z)\big)}{(1-|z|^2)^{2\alpha}}dm(z)<\infty.$$

(ii) *$C_\phi:\mathcal{B}^\alpha\to\mathcal{Q}_p$ exists as a compact operator if and only if $\phi\in\mathcal{Q}_p$ and*

$$\lim_{r\to1}\sup_{w\in\phi(\mathbf{D})}\int_{\phi(\mathbf{D})}\frac{\exp\big(-p\rho_{\phi(\mathbf{D})}(w,z)\big)}{(1-|z|^2)^{2\alpha}}1_{\Delta_r}(z)dm(z)=0.$$

(iii) *$C_\phi:\mathcal{Q}_p\to\mathcal{B}^\alpha$ exists as a bounded operator if and only if*

$$\sup_{w\in\phi(\mathbf{D})}\frac{\delta_{\phi(\mathbf{D})}(w)}{\delta_{\mathbf{D}}(w)}\exp\big((1-\alpha)\rho_{\phi(\mathbf{D})}(\phi(0),w)\big)<\infty.$$

(iv) *$C_\phi:\mathcal{Q}_p\to\mathcal{B}^\alpha$ exists as a compact operator if and only if $\phi\in\mathcal{B}^\alpha$ and*

$$\lim_{r\to1}\sup_{w\in\phi(\mathbf{D})}\frac{\delta_{\phi(\mathbf{D})}(w)}{\delta_{\mathbf{D}}(w)}\exp\big((1-\alpha)\rho_{\phi(\mathbf{D})}(\phi(0),w)\big)1_{\Delta_r}(w)=0.$$

Proof. Let $G = \phi(\mathbf{D})$. Since $\phi : \mathbf{D} \to \mathbf{D}$ is conformal, $a \in \mathbf{D}$ is equivalent to $w = \phi(a) \in G$. If ψ stands for the inverse map of ϕ, then by (2.12)

$$1 - |\sigma_a(\psi(z))|^2 \approx \exp(-\rho_{\mathbf{D}}(0, \sigma_a(\psi(z))) = \exp(-\rho_G(\phi(a), z)),$$

and hence

$$\begin{aligned}\int_{\mathbf{D}} \frac{|\phi'(u)|^2(1 - |\sigma_a(u)|^2)^p}{(1 - |\phi(u)|^2)^{2\alpha}} dm(u) &= \int_G \frac{(1 - |\sigma_a(\psi(z))|^2)^p}{(1 - |z|^2)^{2\alpha}} dm(z)\\ &\approx \int_G \frac{\exp\big(-p\rho_{\phi(\mathbf{D})}(w, z)\big)}{(1 - |z|^2)^{2\alpha}} dm(z).\end{aligned}$$

This, together with Theorem 2.2.1 (i) and (ii), leads to Theorem 3.1.1 (i) and (ii).

Because of (2.12) and (2.13), the above argument, now using Theorem 2.2.1 (iii) and (iv), yields the rest two results in Theorem 2.3.1.

As an application of these theorems, we present an example of a composition operator that distinguishes $\mathcal{Q}_p$ from $\mathcal{B}^\alpha$.

Example 2.3.1. Let $\alpha \in (0, 1]$ and $p \in (0, \infty)$. Then there exists a univalent self-map ϕ of $\mathbf{D}$ such that $C_\phi : \mathcal{B}^\alpha \to \mathcal{Q}_p$ is bounded but $\|\phi\|_{H^\infty} = 1$.

Proof. Set $G_0 = \mathbf{D}(0, 1/2)$, the disk with radius $1/2$ and center 0. For $k \in \mathbf{N}$, choose $a_k \in \mathbf{D}$ such that $\rho_{\mathbf{D}}(0, a_k) \geq k$ and the disks $G_k = \sigma_{a_k}(G_0)$ have disjoint projections onto $\mathbf{T}$. We note for use below that the diameter of G_k is comparable to $1 - |a_k|$. Let S_k be the convex hull of $G_k \cup \{0\}$ and define $G = G_0 \cup \bigcup_{k=1}^\infty S_k$. Then G is a simply connected subdomain of $\mathbf{D}$, and we can take ϕ to be the Riemann map from $\mathbf{D}$ onto G with $\phi(0) = 0$ and $\phi'(0) > 0$. We demonstrate that ϕ is the required map.

Note that $\delta_G(a_k) \to 0$, since $\delta_G(a_k) \leq 1 - |a_k|$ and $\rho_{\mathbf{D}}(0, a_k) \to \infty$. Accordingly, $\|\phi\|_{H^\infty} = 1$.

Next, for $\alpha \in (0, 1]$ and $p \in (0, \infty)$ fixed, we use Theorem 2.3.1 to show that $C_\phi : \mathcal{B}^\alpha \to \mathcal{Q}_p$ is bounded. First, let $w \in G_0$. Since $G_k \subset G$, the comparison principle for the hyperbolic metric yields that $\rho_G \leq \rho_{G_k}$. Then, for $z \in G$, $\exp\big(-p\rho_G(w, z)\big) \preceq (1 - |w|)^p$, by (2.12), since $\rho_{\mathbf{D}}(0, w) \leq \log 3$. Hence

$$\int_{S_k} \frac{\exp\big(-p\rho_G(w, z)\big)}{(1 - |z|^2)^{2\alpha}} dm(z) \preceq \int_{S_k} \frac{dm(z)}{(1 - |z|^2)^{2\alpha - p}} \preceq (1 - |a_k|)^{p - 2\alpha + 2}.$$

Since

$$1 - |a_k| \leq 2\exp(-\rho_{\mathbf{D}}(0, a_k)) \leq 2\exp(-k),$$

it follows that

$$\int_G \frac{\exp\big(-p\rho_G(w, z)\big)}{(1 - |z|^2)^{2\alpha}} dm(z) \preceq \sum_{j=0}^\infty \exp(-j(p - 2\alpha + 2)) \preceq 1$$

for any $w \in G_0$. Next suppose $w \in S_k$. Then, as above,

$$\int_{\cup_{j\neq k} S_j} \frac{\exp\big(-p\rho_G(w,z)\big)}{(1-|z|^2)^{2\alpha}}\,dm(z) \preceq \sum_{j\neq k}^{\infty} \exp(-j(p-2\alpha+2)) \preceq 1,$$

and

$$\int_{S_k} \frac{\exp\big(-p\rho_G(w,z)\big)}{(1-|z|^2)^{2\alpha}}\,dm(z) \preceq \int_0^{|a_k|} \frac{1-|a_k|}{(1-r)^{2\alpha}}\,dr \preceq 1,$$

since $\alpha \in (0,1]$ and $\exp\big(-p\rho_G(w,z)\big) \le 1$. Putting the previous estimates altogether, we get that

$$\sup_{w\in G} \int_G \frac{\exp\big(-p\rho_G(w,z)\big)}{(1-|z|^2)^{2\alpha}}\,dm(z) < \infty,$$

and so that $C_\phi : \mathcal{B}^\alpha \to \mathcal{Q}_p$ is a bounded operator, by Theorem 3.1.1.

Notes

2.1 Theorem 2.1.1 can be reformulated as: there is a holomorphic map $F : \mathbf{D} \to \mathbf{C}^2$ such that $(1-|z|^2)^\alpha |F'(z)| \approx 1$ for all $z \in \mathbf{D}$ (for more information, see also Gauthier-Xiao [69]). In case of the Bloch space (viz. $\alpha = 1$), this theorem is due to Ramey and Ullrich [103]. In [143], Zhao gave some characterizations of $\mathcal{B}^\alpha$ after the fashion of $\mathcal{Q}_p$.

2.2 See the books of Cowen and MacCluer [43] and Shapiro [110], as well as the conference proceeding edited by Jafari et. al [81] for the discussions of composition operators on the classical spaces of holomorphic functions. Theorem 2.2.1 generalizes some results of Smith-Zhao [115], Xiao [137], Madigan [88] and Madigan-Matheson [89]. Very recently, Lindström, Makhmutov and Taskinen [87] proved essentially that in case of $\alpha = 1$, the left-hand sides of (2.5) and (2.6) are comparable with the norm and the essential norm of $C_\phi : \mathcal{B} \to \mathcal{Q}_p$ respectively, and their result improves Montes-Rodriguez's [93]. Meantime, it is worthwhile to remark that (2.7) and (2.8) are independent of $p \in (0,\infty)$, and thus they are the conditions for $C_\phi : \mathcal{B} \to \mathcal{B}^\alpha$ to be bounded and compact respectively (cf. [139]). It would be interesting to characterize the composition operators C_ϕ sending the spaces $\mathcal{Q}_p$ to themselves in terms of the function theoretic properties of ϕ. In the connection with this topic, we would like to mention: Aulaskari-Zhao [28], Bourdon-Cima-Matheson [34], Danikas-Ruscheweyh-Siskakis [47], Lou [86], Smith [114], Tjani[124], Wirths-Xiao [129], Xiao [137] and Zorboska[145].

2.3 For those facts related to the hyperbolic metric, see, for example, Ahlfors [4, Chapter 1] and Shapiro [110, p.157]. The analogs of both Theorem 2.3.1 (i) and Example 2.3.1 for $\alpha = 1$ was established earlier in Smith-Zhao [115].

3. Series Expansions

We saw in the previous two chapters that the series with Hadamard gaps play an important role in studying $\mathcal{Q}_p$. Accordingly, it is necessary to investigate the properties of the coefficients of the usual or random power series living on $\mathcal{Q}_p$ in detail.

3.1 Power Series

Recall that if $f(z) = \sum_{n=0} a_n z^n$ and $p \in [0, \infty)$, then

$$\|f\|^2_{\mathcal{D}_p} \approx \sum_{n=1}^{\infty} n^{1-p}|a_n|^2. \tag{3.1}$$

However, regarding $\mathcal{Q}_p$, we have the following result which is quite complicated.

Theorem 3.1.1. *Let $p \in (0, \infty)$ and $f \in \mathcal{H}$ with $f(z) = \sum_{n=0}^{\infty} a_n z^n$. Then $f \in \mathcal{Q}_p$ if and only if*

$$\sup_{w \in \mathbf{D}} \sum_{n=0}^{\infty} \frac{(1-|w|^2)^p}{(n+1)^{p+1}} \left| \sum_{m=0}^{n} \frac{(m+1)a_{m+1}\Gamma(n-m+p)}{(n-m)!} \bar{w}^{n-m} \right|^2 < \infty.$$

Proof. Under the assumption above, we have

$$\begin{aligned} \frac{f'(z)}{(1-\bar{w}z)^p} &= \sum_{n=0}^{\infty}(n+1)a_{n+1}z^n \sum_{n=0}^{\infty} \frac{\Gamma(n+p)}{n!\Gamma(p)} \bar{w}^n z^n \\ &= \sum_{n=0}^{\infty} \left(\sum_{m=0}^{n} \frac{(m+1)a_{m+1}\Gamma(n-m+p)}{(n-m)!\Gamma(p)} \bar{w}^{n-m} \right) z^n \end{aligned}$$

This, together with Parseval's formula, implies

$$\begin{aligned} \frac{F_p(f,w)}{(1-|w|^2)^p} &= \int_{\mathbf{D}} |f'(z)|^2 \frac{(1-|z|^2)^p}{|1-\bar{w}z|^{2p}} dm(z) \\ &= \int_{\mathbf{D}} \left| \sum_{n=0}^{\infty} \left(\sum_{m=0}^{n} \frac{(m+1)a_{m+1}\Gamma(n-m+p)}{(n-m)!\Gamma(p)\bar{w}^{m-n}} \right) z^n \right|^2 \frac{dm(z)}{(1-|z|^2)^{-p}} \end{aligned}$$

$$= 2\pi \int_0^1 \sum_{n=0}^{\infty} \left| \sum_{m=0}^{n} \frac{(m+1)a_{m+1}\Gamma(n-m+p)}{(n-m)!\Gamma(p)\bar{w}^{n-m}} \right|^2 \frac{r^{2n}dr}{(1-r^2)^{-p}}$$

$$= \pi \sum_{n=0}^{\infty} \left| \sum_{k=0}^{n} \frac{(m+1)a_{k+1}\Gamma(n-m+p)}{(n-m)!\Gamma(p)} \bar{w}^{n-m} \right|^2 \int_0^1 (1-r)^p r^n dr$$

$$= \pi \sum_{n=0}^{\infty} B(n+1, p+1) \left| \sum_{m=0}^{n} \frac{(m+1)a_{m+1}\Gamma(n-m+p)}{(n-m)!\Gamma(p)} \bar{w}^{n-m} \right|^2 .$$

Note that the Beta function $B(\cdot,\cdot)$ ensures

$$B(n+1,p+1) = \frac{\Gamma(n+1)\Gamma(p+1)}{\Gamma(n+p+2)} \approx \frac{\Gamma(p+1)}{(n+1)^{p+1}},$$

which comes from a simple application of Stirling's formula. So, the equivalence stated in Theorem 3.1.1 follows from Theorem 1.1.1 and the preceding calculation.

Corollary 3.1.1. *Let $p \in (0,\infty)$ and $f \in \mathcal{H}$ with $f(z) = \sum_{n=0}^{\infty} a_n z^n$.*
(i) *The condition*

$$\sup_{0 \le r < 1} \sum_{n=0}^{\infty} \frac{(1-r^2)^p}{(n+1)^{p+1}} \left(\sum_{m=0}^{n} \frac{(m+1)|a_{m+1}|}{(n-m+1)^{1-p}} r^{n-m} \right)^2 < \infty \tag{3.2}$$

implies that $f \in \mathcal{Q}_p$.
(ii) *If $a_n \ge 0$ for $n \in \mathbf{N} \cup \{0\}$ and $f \in \mathcal{Q}_p$, then (3.2) holds.*
(iii) *If $g(z) = \sum_{n=0}^{\infty} b_n z^n$ and $b_n \succeq |a_n|$ for $n \in \mathbf{N}\cup\{0\}$ then $\|g\|_{\mathcal{Q}_p} \succeq \|f\|_{\mathcal{Q}_p}$.*

Proof. The argument for Theorem 3.1.1 implies actually that for $w \in \mathbf{D}$,

$$F_p(f,w) \preceq \sum_{n=0}^{\infty} \frac{(1-|w|^2)^p}{(n+1)^{p+1}} \left| \sum_{m=0}^{n} \frac{(m+1)a_{m+1}\Gamma(n-m+p)}{(n-m)!} |w|^{n-m} \right|^2 ,$$

which yields (i), due to Theorem 1.1.1. Conversely, if $a_n \ge 0$ then it is easy to get (3.2) from $f \in \mathcal{Q}_p$. Now, (iii) is a consequence of (i) and (ii).

Remark that the above assertion (iii) is particularly useful when one wishes to determine the $\mathcal{Q}_p$-membership of a power series whose Taylor coefficients decay rapidly at the infinity. For example, any $f \in \mathcal{H}$ with $f(z) = \sum_{n=0}^{\infty} a_n z^n$; $|a_n| \preceq n^{-1}$, must belong to $\mathcal{Q}_p$, $p > 0$, in that so does $\log(1-z)$.

3.2 Partial Sums

The partial sums of power series can be selected to characterize every individual function in $\mathcal{Q}_p$.

Let $f(z) = \sum_{k=0}^{\infty} a_k z^k$ be in $\mathcal{H}$. For $n \in \mathbf{N} \cup \{0\}$, define

$$s_n(f)(z) = \sum_{k=0}^{n} a_k z^k$$

and

$$\sigma_n(f)(z) = \sum_{k=0}^{n} \frac{s_k(f)(z)}{n+1} = \sum_{k=0}^{n} \left(1 - \frac{k}{n+1}\right) a_k z^k.$$

It is known that

$$\sigma_n(f)(z) = \frac{1}{2\pi} \int_{\mathbf{T}} K_n(\zeta) f(z\zeta) |d\zeta|,$$

where

$$K_n(\zeta) = \sum_{k=-n}^{n} \left(1 - \frac{|k|}{n+1}\right) \bar{\zeta}^k, \quad \zeta \in \mathbf{T},$$

is the Fejér kernel with the identical property:

$$1 = \frac{1}{2\pi} \int_{\mathbf{T}} K_n(\zeta) |d\zeta|.$$

Theorem 3.2.1. *Let $p \in [0, \infty)$ and let $f \in \mathcal{H}$. Then $f \in Q_p$ if and only if*

$$\sup\{\|\sigma_n(f)\|_{Q_p} : n \in \mathbf{N} \cup \{0\}\} < \infty.$$

Proof. If $f \in \mathcal{Q}_p$, then

$$(\sigma_n(f))'(z) = \frac{1}{2\pi} \int_{\mathbf{T}} K_n(\zeta) f'(z\zeta) \zeta |d\zeta|.$$

Putting $f_\zeta(z) = f(\zeta z)$ and using Minkowski's inequality, we see that for any $w \in \mathbf{D}$,

$$F_p(\sigma_n(f), w) \le \frac{1}{2\pi} \int_{\mathbf{T}} F_p(f_\zeta, w) K_n(\zeta) |d\zeta| \le \|f\|_{\mathcal{Q}_p}.$$

Applying Theorem 1.1.1, we find that $\sigma_n(f) \in \mathcal{Q}_p$ with $\|\sigma_n(f)\|_{\mathcal{Q}_p} \le \|f\|_{\mathcal{Q}_p}$. Conversely, let $\sup_{n \in \mathbf{N} \cup \{0\}} \|\sigma_n(f)\|_{\mathcal{Q}_p}$ be finite. Noticing

$$f_r(z) = f(rz) = (1-r)^2 \sum_{n=0}^{\infty} (n+1) \sigma_n(f)(z) r^n, \quad r \in (0,1),$$

we have that for $w \in \mathbf{D}$,

$$F_p(f_r, w) \le (1-r)^2 \sum_{n=0}^{\infty} (n+1) F_p(\sigma_n(f), w) r^n \le \sup_{n \in \mathbf{N} \cup \{0\}} \|\sigma_n(f)\|_{\mathcal{Q}_p}.$$

By Fatou's lemma, we obtain that $F_p(f, w) \le \sup_{n \in \mathbf{N} \cup \{0\}} \|\sigma_n(f)\|_{\mathcal{Q}_p}$. Hence $f \in \mathcal{Q}_p$ by Theorem 1.1.1.

3.3 Nonnegative Coefficients

Although the quintessential example $-\log(1-z)$ does not distinguish each $\mathcal{Q}_p$, its special expansion $\sum_{n=1}^{\infty} n^{-1}z^n$ suggests us to consider the question of characterizing those $f \in \mathcal{Q}_p$ for which $f(z) = \sum_{n=0}^{\infty} a_n z^n$ with $a_n \geq 0$.

Theorem 3.3.1. *Let $f \in \mathcal{H}$ with $f(z) = \sum_{n=0}^{\infty} a_n z^n$ and $a_n \geq 0$ for $n \in \mathbf{N} \cup \{0\}$.*

(i) If $p \in (0,1]$ then $f \in \mathcal{Q}_p$ if and only if

$$\sup_{k\in\mathbf{N}} k^{-p} \sum_{n=0}^{\infty} (n+1)^{1-p} \left(\sum_{m=0}^{\min(n,k)} \frac{a_{2n-m+1}}{(m+1)^{1-p}} \right)^2 < \infty. \tag{3.3}$$

(ii) If $p \in (1,\infty)$ then $f \in \mathcal{Q}_p$ if and only if $\sup_{n\in\mathbf{N}} n^{-1} \sum_{k=1}^{n} k a_k < \infty$.

Proof. (i) From the proof of Theorem 3.1.1 it follows that for $w \in \mathbf{D}$,

$$\frac{f'(z)}{(1-\bar{w}z)^p} = \sum_{n=0}^{\infty} (n+1)c_{n+1} z^n = \Big(\sum_{n=1}^{\infty} c_n z^n \Big)',$$

where

$$(n+1)c_{n+1} = \sum_{n=0}^{\infty} \frac{(m+1)a_{m+1}\Gamma(n-m+p)}{(n-m)!\Gamma(p)} \bar{w}^{n-m}.$$

Now invoking (3.1), we get

$$\begin{aligned} \frac{F_p(f,w)}{(1-|w|^2)^p} &\approx \sum_{n=0}^{\infty} (n+1)^{1-p} |c_{n+1}|^2 \\ &\preceq \sum_{n=0}^{\infty} (n+1)^{1-p} \left(\sum_{m=0}^{n} \frac{(m+1)a_{m+1}|w|^{n-m}}{(n+1)(n-m+1)^{1-p}} \right)^2. \end{aligned}$$

Let k be the positive integer satisfying: $(k+1)^{-1} < 1-|w| \leq k^{-1}$. Using the above inequalities, we see that it suffices to verify

$$\sup_{k\in\mathbf{N}} k^{-p} I_k < \infty, \tag{3.4}$$

where, for each $k \in \mathbf{N}$, one sets

$$I_k = \sum_{n=0}^{\infty} (n+1)^{1-p} \left(\sum_{m=0}^{n} \frac{(m+1)a_{m+1}(1-(k+1)^{-1})^{n-m}}{(n+1)(n-m+1)^{1-p}} \right)^2. \tag{3.5}$$

We assume now that the sequence of coefficients a_n of f obeys (3.3). Then

$$\sum_{n=k}^{2k}(n+1)^{1-p}\left(\sum_{m=0}^{n}\frac{a_{2n-m+1}}{(m+1)^{1-p}}\right)^2 \preceq k^p, \quad k\in\mathbf{N},$$

and hence

$$\sum_{n=k}^{2k}\Big(\sum_{m=n}^{2n}|a_m|\Big)^2 \preceq k, \quad k\in\mathbf{N}.$$

This estimation gives

$$\sum_{m=n}^{2n} a_m \preceq 1, \quad n\in\mathbf{N},$$

and consequently,

$$\sum_{m=1}^{n} m|a_m| \preceq n, \quad n\in\mathbf{N}. \tag{3.6}$$

Using (3.6) we can simplify (3.4) by observing that for $k\in\mathbf{N}$,

$$\sum_{n=0}^{\infty}(n+1)^{1-p}\left(\sum_{0\le m\le n/2}\frac{(m+1)a_{m+1}(1-(k+1)^{-1})^{n-m}}{(n+1)(n-m+1)^{1-p}}\right)^2 \preceq k^p$$

due to the binomial theorem. It now follows that we need only to prove that (3.3) implies

$$\sum_{n=0}^{\infty}(n+1)^{1-p}\left(\sum_{0\le m\le n/2}\frac{a_{n-m+1}(1-k^{-1})^m}{(m+1)^{1-p}}\right)^2 \preceq k^p$$

or, more simply, $J_k \preceq k^p$ for $k\in\mathbf{N}$, where

$$J_k=\sum_{n=0}^{\infty}(n+1)^{1-p}\left(\sum_{m=0}^{n}\frac{a_{2n-m+1}(1-k^{-1})^m}{(m+1)^{1-p}}\right)^2. \tag{3.7}$$

As our last reduction we first notice that for $0\le m\le k$ we have $(1-k^{-1})^m\approx 1$. Fixing a large integer N, then splitting the sum in (3.7) into two parts and applying (3.3), we obtain

$$\begin{aligned}\frac{J_k}{2} &\le \sum_{n=0}^{\infty}(n+1)^{1-p}\left(\sum_{m=0}^{\min(n,kN)}\frac{a_{2n-m+1}}{(m+1)^{1-p}}\right)^2 \\ &\quad + \sum_{n=kN}^{\infty}(n+1)^{1-p}\left(\sum_{m=kN}^{n}\frac{a_{2n-m+1}(1-k^{-1})^m}{(m+1)^{1-p}}\right)^2 \\ &\preceq (kN)^p+\left(\frac{1-k^{-1}}{1-(kN)^{-1}}\right)^{2kN} J_{kN}.\end{aligned}$$

Thus

$$\sup_{k\in\mathbf{N}} \frac{J_k}{k^p} \preceq N^p + \sup_{k\in\mathbf{N}} \frac{J_{kN}}{(kN)^p}$$

provided N is sufficiently large. This yields $\sup_{k\in\mathbf{N}} J_k k^{-p} \preceq N^p$ provided f is a polynomial. A limit argument concludes the proof of sufficiency part of (i).

The proof of necessity part of (i) is easy. By reversing the first step in the above proof, we obtain

$$(1-|w|^2)^p \sum_{n=0}^{\infty} (n+1)^{1-p} \left(\sum_{m=0}^{n} \frac{(m+1)a_{m+1}|w|^{n-m}}{(n+1)(n-m+1)^{1-p}} \right)^2 \preceq \|f\|^2_{\mathcal{Q}_p}$$

for all $w \in \mathbf{D}$. Now (3.3) is established by replacing $1-|w|$ with k^{-1} and $|w|^{n-m}$ with 1 provided $n-m \le k$. The remaining terms can be ignored.

(ii) In case of $p \in (1,\infty)$, one has $\mathcal{Q}_p = \mathcal{B}$. So, if $f \in \mathcal{Q}_p$, then for $j \in \mathbf{N}$,

$$\begin{aligned}
\|f\|_{\mathcal{B}} &\ge \sup_{z=1-j^{-1}} (1-|z|^2)|f'(z)| \\
&\ge j^{-1} \sum_{n=1}^{j} n a_n (1-j^{-1})^{n-1} \\
&\ge j^{-1}(1-j^{-1})^{j-1} \sum_{n=1}^{j} n a_n \\
&\succeq j^{-1} \sum_{n=1}^{j} n a_n,
\end{aligned}$$

and hence $\sup_{j\in\mathbf{N}} j^{-1} \sum_{n=1}^{j} n a_n$ is finite.

On the other hand, if $\sup_{j\in\mathbf{N}} j^{-1} \sum_{n=1}^{j} n a_n < \infty$, then

$$\sum_{n=2^k}^{2^{k+1}} a_n \preceq 1, \quad k \in \mathbf{N},$$

and hence for $z \in \mathbf{D}$,

$$\begin{aligned}
|f'(z)| &= \left| \sum_{k=0}^{\infty} \sum_{n=2^k}^{2^{k+1}-1} n a_n z^{n-1} \right| \\
&\le \sum_{k=0}^{\infty} 2^{k+1} \sum_{n=2^k}^{2^{k+1}-1} a_n |z|^{2^k-1} \\
&\preceq \sum_{k=0}^{\infty} 2^k |z|^{2^k-1} \\
&\preceq (1-|z|)^{-1},
\end{aligned}$$

which implies $f \in \mathcal{Q}_p$. The proof is finished.

As a direct consequence of Theorem 3.3.1, the following conclusion supplies us with a surprising reason why $\log(1-z)$ lies in each $\mathcal{Q}_p$.

Corollary 3.3.1. *Let $p \in (0,\infty)$ and let $f(z)=\sum_{n=0}^{\infty} a_n z^n$ with a_n nonnegative and nonincreasing. Then $f \in \mathcal{Q}_p$ if and only if $\sup_{n\in\mathbf{N}} na_n < \infty$.*

Proof. For convenience, let $C=\sup_{n\in\mathbf{N}} na_n$.

Case 1: $p\in(0,1]$. Suppose that $f\in\mathcal{Q}_p$ and

$$S(k)=\sum_{n=0}^{\infty}(n+1)^{1-p}\left(\sum_{m=0}^{\min(n,k)}\frac{a_{2n-m+1}}{(m+1)^{1-p}}\right)^2,$$

for $k\in\mathbf{N}$. Using the assumption that $\{a_n\}$ is a nonnegative and nonincreasing sequence, we obtain

$$\begin{aligned} S(k) &\ge \sum_{n=0}^{k}(n+1)^{1-p}\left(\sum_{m=0}^{\min(n,k)}\frac{a_{2n-m+1}}{(m+1)^{1-p}}\right)^2\\ &\ge \sum_{n=0}^{k}(n+1)^{1-p}\left(\sum_{m=0}^{n}\frac{a_{2n+1}}{(m+1)^{1-p}}\right)^2\\ &\ge a_{2k+1}^2\sum_{n=0}^{k}(n+1)^{1+p}\\ &\succeq k^{p+2}a_{2k+1}^2, \end{aligned}$$

and so $C<\infty$, by Theorem 3.3.1 (i).

Conversely, under the condition that C is finite, we dominate the upper bound of $S(k)$ as follows:

$$\begin{aligned} S(k) &= \left(\sum_{n=0}^{k}+\sum_{n=k+1}^{\infty}\right)(n+1)^{1-p}\left(\sum_{m=0}^{\min(n,k)}\frac{a_{2n-m+1}}{(m+1)^{1-p}}\right)^2\\ &\le \sum_{n=0}^{k}(n+1)^{1-p}\left(\sum_{m=0}^{n}\frac{a_{n+1}}{(m+1)^{1-p}}\right)^2\\ &+\sum_{n=k+1}^{\infty}(n+1)^{1-p}\left(\sum_{m=0}^{k}\frac{a_{2n-k+1}}{(m+1)^{1-p}}\right)^2\\ &\preceq C^2\sum_{n=0}^{k}(n+1)^{-1-p}\left(\sum_{m=0}^{n}\frac{1}{(m+1)^{1-p}}\right)^2\\ &+C^2\sum_{n=k+1}^{\infty}\frac{(n+1)^{1-p}}{(2n-k+1)^2}\left(\sum_{m=0}^{k}\frac{1}{(m+1)^{1-p}}\right)^2 \end{aligned}$$

$$
\begin{aligned}
&\preceq C^2\left(\sum_{n=0}^{k}(n+1)^{p-1}+\sum_{n=k+1}^{\infty}\frac{(n+1)^{1-p}}{(2n-k+1)^2}\right)\\
&\preceq C^2\left(k^p+\sum_{n=k+1}^{\infty}\frac{1}{(n+1)^{1+p}}\right)\\
&\preceq C^2k^p.
\end{aligned}
$$

This gives that $k^{-p}S(k)\preceq C^2$ and so that $f\in\mathcal{Q}_p$, by Theorem 3.3.1 (i).

Case 2: $p\in(1,\infty)$. At this point, we have $\mathcal{Q}_p=\mathcal{B}$. Note that under the hypothesis of Corollary 3.3.1, one has

$$\frac{na_n}{2}\le n^{-1}\sum_{k=1}^{n}ka_k\le C,\quad n\in\mathbf{N}.$$

Thus the desired result follows from Theorem 3.3.1 (ii).

Another direct consequence of Theorem 3.3.1 is the construction of some special functions suggesting that $\mathcal{Q}_p$ is a large subspace of $\mathcal{D}_p$.

Example 3.3.1. Let $p\in(0,1)$. Then there exists $f\in\mathcal{D}_p\cap\left(\cap_{p<q}\mathcal{Q}_q\right)\setminus\mathcal{Q}_p$

Proof. For $a_j=(2^{j(1-p)}j^{2(1+\beta)})^{-1/2}$, $j\in\mathbf{N}$, and $\beta\in(0,p/2)$, let

$$f_\beta(z)=\sum_{j=1}^{\infty}a_j\sum_{m=0}^{j}z^{m+2^j}=\sum_{n\ge1}b_nz^n.$$

Step 1: we first show $f_\beta\in\mathcal{D}_p$. Since

$$\|f_\beta\|_{\mathcal{D}_p}^2=\sum_{j=1}^{\infty}|a_j|^2\sum_{m=0}^{j}(m+2^j)^{1-p}\approx\sum_{j=1}^{\infty}j^{-(1+2\beta)}<\infty.$$

Step 2: we next show $f_\beta\in\cap_{p<q}\mathcal{Q}_q$. By Corollary 2.2.1, it suffices to prove $f_\beta\in\mathcal{B}^{(1+p)/2}$. Suppose $2^{j-1}\le n<2^j$, then

$$\sum_{m=n}^{2n}b_m\le\sum_{m=2^{j-1}}^{2^{j+1}}b_m\preceq n^{-(1-p)/2},$$

and hence

$$
\begin{aligned}
|f'_\beta(z)|&=\left|\sum_{k=0}^{\infty}\sum_{n=2^k}^{2^{k+1}-1}nb_nz^{n-1}\right|\\
&\le\sum_{k=0}^{\infty}2^{k+1}\sum_{n=2^k}^{2^{k+1}-1}b_n|z|^{2^k-1}
\end{aligned}
$$

$$\preceq \sum_{k=0}^{\infty} 2^{(1+p)k/2}|z|^{2^k-1}$$

$$\preceq \sum_{j=0}^{\infty}(1+j)^{-(1-p)/2}|z|^j$$

$$\preceq \frac{1}{(1-|z|)^{(1+p)/2}},$$

which implies $f_\beta \in \mathcal{B}^{(1+p)/2}$.

Step 3: we finally show $f_\beta \notin \mathcal{Q}_p$ as $\beta < p/2$. By Theorem 3.3.1 (i) it suffices to show

$$\sup_{k\in\mathbf{N}} k^{-p} \sum_{n=2^{2k}}^{\infty} n^{1-p}\left(\sum_{m=0}^{k} \frac{b_{2n-m+1}}{(m+1)^{1-p}}\right)^2 = \infty. \tag{3.8}$$

Let $j \geq 2k$ and observe that the interval $2^j + j/4 \leq n \leq 2^j + j/2$ contains $[j/4]$ (the integer part of $j/4$) integers each of which satisfies $2^{j+1} \leq 2n - m + 1 \leq 2^{j+1} + j + 1$ and hence

$$\sum_{n=2^j}^{2^{j+1}-1} n^{1-p}\left(\sum_{m=0}^{k} \frac{b_{2n-m+1}}{(m+1)^{1-p}}\right)^2 \succeq (2^j)^{1-p} a_{j+1}^2 j\left(\sum_{m=0}^{k}(1+m)^{p-1}\right)^2$$

$$\succeq \frac{k^{2p}}{j^{1+2\beta}}.$$

Consequently

$$\frac{1}{k^p}\sum_{n=2^{2k}}^{\infty} n^{1-p}\left(\sum_{m=0}^{k} \frac{b_{2n-m+1}}{(m+1)^{1-p}}\right)^2 = \frac{1}{k^p}\sum_{j=2k}^{\infty}\sum_{n=2^j}^{2^{j+1}-1} n^{1-p}\left(\sum_{m=0}^{k} \frac{b_{2n-m+1}}{(m+1)^{1-p}}\right)^2$$

$$\succeq \sum_{j=2k}^{\infty} \frac{k^p}{j^{1+2\beta}} \approx k^{p-2\beta} \to \infty$$

provided $\beta < p/2$. Thus (3.8) follows and the proof is complete.

3.4 Random Series

Now let $\varepsilon_n(\omega)$ be a Bernoulli sequence of random variables on a probability space. In other words, the random variables are independent and each ϵ_n takes the values 1 and -1 with equal probability 1/2. If $f(z) = \sum_{n=0}^{\infty} a_n z^n$ is in $\mathcal{H}$, then we let $f_\omega(z) = \sum_{n=0}^{\infty} \epsilon_n a_n z^n$, and call f_ω the random series of f. Moreover, a.s. means "almost surely"; that is, "for almost every choice of signs" .

Theorem 3.4.1. *Let $p \in [0,1)$ and $f \in \mathcal{H}$ with $f(z) = \sum_{n=0}^{\infty} a_n z^n$. If $f \in \mathcal{Q}_p$ then $\sum_{n=0}^{\infty} n^{1-p}|a_n|^2 < \infty$. Conversely, if $\sum_{n=0}^{\infty} n^{1-p}|a_n|^2 < \infty$ then $f_\omega \in \mathcal{Q}_p$ a.s..*

Proof. The first part of the theorem is trivial since $\mathcal{Q}_p \subseteq \mathcal{D}_p$. However, the second part is surprising. When $\sum_{n=1}^{\infty} n^{1-p}|a_n|^2$ is finite, we have $f \in \mathcal{D}_p$. By Theorem 2 in [42] f_ω is a pointwise multiplier of $\mathcal{D}_p$, that is, $f_\omega g \in \mathcal{D}_p$ for any $g \in \mathcal{D}_p$. Upon taking

$$g(z) = \left(\frac{1-|w|^2}{1-\bar{w}z}\right)^{p/2}, \quad w \in \mathbf{D},$$

and applying Theorem 1.1(c) in [117], we get that $\|g\|_{\mathcal{D}_p} \preceq 1$ (cf. Lemma 1.4.1) and

$$\int_{\mathbf{D}} |g'(z)|^2(1-|z|^2)^p |f_\omega'(z)|^2 dm(z) \preceq \|g\|_{\mathcal{D}_p}^2 \preceq 1,$$

which certainly implies $f_\omega \in \mathcal{Q}_p$. We are done.

Next we show that $\sum_{n=0}^{\infty} n^{1-p}|a_n|^2 < \infty$ is best possible in a very strong sense.

Theorem 3.4.2. *Let $p \in [0,1)$ and $f \in \mathcal{H}$ with $f(z) = \sum_{n=0}^{\infty} a_n z^n$. Given a sequence $\{c_n\}$, $c_n \searrow 0$, one can choose coefficients $a_n > 0$ such that $\sum_{n=0}^{\infty} n^{1-p} a_n^2 c_n < \infty$ but $f_\omega \notin \mathcal{Q}_p$ for any choice of ω.*

Proof. Let $\{c_n\}$ be a sequence of positive constants decreasing monotonically to 0. Choose integers n_k which satisfy:

$$(i)\ n_0 = 1,$$

$$(ii)\ n_k > 2n_{k-1}, \quad k \in \mathbf{N},$$

$$(iii)\ \sum_{k=0}^{\infty} c_{n_k}^{1/2} < \infty.$$

Define $a_n > 0$ by $a_1 = 1$ and $n^{1-p}a_n^2 = n_k^{-1} c_{n_{k-1}}^{-1/2}$ for $n_{k-1} \le n < n_k$. Then

$$\begin{aligned}
\sum_{n=1}^{\infty} n^{1-p} a_n^2 c_n &= \sum_{k=1}^{\infty} \sum_{n=n_{k-1}}^{n_k - 1} n^{1-p} a_n^2 c_n \\
&\le \sum_{k=1}^{\infty} \sum_{n=n_{k-1}}^{n_k - 1} n_k^{-1} c_{n_{k-1}}^{-1/2} c_{n_{k-1}} \\
&= \sum_{k=1}^{\infty} (n_k - n_{k-1}) n_k^{-1} c_{n_{k-1}}^{1/2} \\
&\le \sum_{k=1}^{\infty} c_{n_{k-1}}^{1/2} < \infty.
\end{aligned}$$

On the other hand,

$$\sum_{n=1}^{\infty} n^{1-p} a_n^2 = \sum_{k=1}^{\infty} \sum_{n=n_{k-1}}^{n_k - 1} n_k^{1-p} c_{n_{k-1}}^2 \geq \frac{1}{2} \sum c_{n_{k-1}}^{-1/2} = \infty,$$

thus $f_\omega \notin \mathcal{D}_p$ and hence $f_\omega \notin \mathcal{Q}_p$ for any choice of ω. The theorem is proved.

Notes

3.1 Theorem 3.1.1 and its proof (due to Aulaskari, Girela and Wulan [18]), are motivated by Proposition of Aulaskari-Xiao-Zhao [27]. Meanwhile, Corollary 1.1.1 (i) can be extended to the Hadamard product. More precisely, for $f(z) = \sum_{n=0}^{\infty} a_n z^n$ and $g(z) = \sum_{n=0}^{\infty} b_n z^n$, we define the Hadamard product of f and g as $f * g(z) = \sum_{n=0}^{\infty} a_n b_n z^n$ (see, for example, Anderson-Clunie-Pommerenke [8]). If $p, q \in [0, \infty)$ and $f \in \mathcal{H}$, then by (3.1) it follows that $f \in \mathcal{Q}_p$ if and only if

$$\sup_{w \in \mathbf{D}} \|(f \circ \sigma_w - f(w)) * k_{p,q}\|_{\mathcal{D}_q} < \infty,$$

where

$$k_{p,q}(z) = \sum_{n=0}^{\infty} n^{(q-p)/2} z^n.$$

See also Aulaskari-Girela-Wulan [17]. Moreover, Corollary 3.1.1 has been used by Aulaskari-Girela-Wulan [18] to prove that if the Taylor coefficients a_n of $f \in \mathcal{Q}_p$ are nonnegative, and if $g \in \mathcal{B}$, then $f * g \in \mathcal{Q}_p$. Concerning the other types of algebraic properties of $\mathcal{Q}_p$, we refer the reader to Aulaskari-Danikas-Zhao [16].

3.2 For the versions of $\mathcal{B}$ and $BMOA$ of Theorem 3.2.1, see Holland-Walsh [77].

3.3 Theorem 3.3.1 (i) is taken from Aulaskari-Stegenga-Xiao [24]. For $BMOA$, this is a well-known unpublished result of Fefferman, see for example [113]. Theorem 3.3.1 (ii) is from Girela [70]. For more general topics on the Taylor coefficients of functions in $\mathcal{B}$, see Anderson-Clunie-Pommerenke [8] and Bennett-Stegenga-Timoney [32]. Corollary 3.3.1 has its root in [33] where Bergh dealt with those BMO and $\mathcal{B}$ functions with nonnegative Taylor coefficients in an approach based on the Fourier series. Of course, Corollary 3.3.1 corresponds nicely to Corollary 1.3.1 which displays a close relation among $\mathcal{B}$, $BMOA$ and $\mathcal{Q}_p$, $p \in (0,1)$ through the value distribution of holomorphic functions under consideration.

Example 3.3.1 should be compared with a constructive example of Holland-Twomey [75]. For the reader's convenience, we give the construction of the example in the sequel. Given $k \in \mathbf{N}$, let $m(k)$ be the integer part of $2^{\sqrt{k}} - 2^{\sqrt{k-1}}$, put

$$F_k = \{2^k + j : j = 0, 1, \cdots, m(k)\},$$

where $F_0 = \{0\}$, and set

$$E_k = \{2^k + j : j \in \cup_{0 \leq l \leq k-1} F_l\},$$

where $E_0 = \{1\}$. Define $a_n = 2^{-\sqrt{k}}$ resp. $a_n = 0$ whenever $n \in E_k$ resp. $n \notin \cup_{j\geq 0} E_j$. If f is the function determined by

$$f(z) = \sum_{n=0}^{\infty} a_n z^n = \sum_{k\geq 0} 2^{-\sqrt{k}} \sum_{n\in E_k} z^n, \quad z \in \mathbf{D},$$

then a further application of Theorem 3.3.1 (cf. [70, Theorem 9.13] for details) gives that

$$f \in \mathcal{B} \cap \left(\cap_{0<q<\infty} H^q \right) \setminus BMOA,$$

where H^q stands for the Hardy space on $\mathbf{T}$ (see also Section 4.2 of next chapter for the definition of H^q).

3.4 Theorems 3.4.1 and 3.4.2 are from Aulaskari-Stegenga-Zhao [25]. According to Theorems 3.4.1 and 1.2.1, random series are similar to lacunary series in case of $\mathcal{Q}_p$, $p \in [0,1)$: they are very well behaved if the coefficients are weightedly square-summable and very badly behaved if not. However, for $p = 1$ Theorem 3.4.1 fails, as shown by Sledd and Stegenga [113]. In case of $\mathcal{B}$, it is easy to figure out that Theorem 3.4.1 is false, see Duren [51] and Sledd [112]. The proof of Theorem 3.4.2 is analogous to that of Theorem 3(b) of Cohran-Shapiro-Ullrich [42].

4. Modified Carleson Measures

In this chapter, we show that $\mathcal{Q}_p$ can be equivalently characterized by means of a modified Carleson measure. In the subsequent three sections, this geometric characterization is used to compare $\mathcal{Q}_p$ with the class of mean Lipschitz functions as well as the Besov space (as one of representatives of the conformally invariant classes of holomorphic functions), and to discuss the mean growth of the derivatives of functions in $\mathcal{Q}_p$.

4.1 An Integral Form

For $p \in (0,\infty)$ we say that a complex Borel measure μ given on $\mathbf{D}$ is a p-Carleson measure provided

$$\|\mu\|_{C_p} = \sup_{I \subseteq \mathbf{T}} \frac{|\mu|(S(I))}{|I|^p} < \infty,$$

where the supremum is taken over all arcs $I \subseteq \mathbf{T}$. Here and elsewhere in the forthcoming chapters, $|I|$ stands for the arclength of I, and that $S(I)$ means the Carleson box based on I:

$$S(I) = \{z \in \mathbf{D} : 1 - |I| \le |z| < 1,\ \frac{z}{|z|} \in I\}.$$

Note that $0 \in S(I)$ if and only if $|I| \ge 1$. So we will always take $|I| < 1$ for granted (unless a special remark is made). When $p = 1$, we get the standard definition of the original Carleson measure. As in [66, p. 239], any p-Carleson measure has an integral representation.

Lemma 4.1.1. *Let $p \in (0,\infty)$ and let μ be a complex Borel measure on $\mathbf{D}$. Then μ is a p-Carleson measure if and only if*

$$|\!|\!|\mu|\!|\!|_{C_p} = \sup_{w \in \mathbf{D}} \int_{\mathbf{D}} \left(\frac{1-|w|}{|1-\bar{w}z|^2}\right)^p d|\mu|(z) < \infty.$$

Proof. Suppose that $|\!|\!|\mu|\!|\!|_{C_p} < \infty$. Then, for the Carleson box $S(I) = \{z \in \mathbf{D} : 1 - h \le |z| < 1,\ |\theta - \arg z| \le h\}$ with $h = |I|$, we take $w = (1-h)e^{i(\theta + h/2)}$, and so have

$$\|\mu\|_{C_p} \geq \inf_{z\in S(I)}\left(\frac{1-|w|}{|1-\bar{w}z|^2}\right)^p |\mu|(S(I)) \succeq \frac{|\mu|(S(I))}{|I|^p},$$

which implies $\|\mu\|_{C_p} \preceq \|\mu\|_{C_p} < \infty$.

Conversely, assume that μ is a p-Carleson measure, that is, $\|\mu\|_{C_p} < \infty$. If $w \in \mathbf{D}(0,3/4)$, then

$$\int_{\mathbf{D}}\left(\frac{1-|w|}{|1-\bar{w}z|^2}\right)^p d|\mu|(z) \preceq |\mu|(\mathbf{D}) \preceq \|\mu\|_{C_p}.$$

If $w \in \mathbf{D}\setminus\mathbf{D}(0,3/4)$, then we put $\mathbf{E}_n = \{z\in\mathbf{D} : |z - w/|w|| < 2^n(1-|w|)\}$ and hence get $|\mu|(\mathbf{E}_n) \preceq \|\mu\|_{C_p}2^{np}(1-|w|)^p$ for $n \in \mathbf{N}$. We also have

$$\frac{1-|w|^2}{|1-\bar{w}z|^2} \preceq \frac{1}{1-|w|}, \quad z\in\mathbf{E}_1,$$

and so for $n \geq 1$ and $\mathbf{E}_0 = \emptyset$,

$$\frac{1-|w|^2}{|1-\bar{w}z|^2} \preceq \frac{1}{2^{2n}(1-|w|)}, \quad z\in\mathbf{E}_n\setminus\mathbf{E}_{n-1}.$$

Consequently,

$$\begin{aligned}\|\mu\|_{C_p} &\leq \sup_{w\in\mathbf{D}}\sum_{n=1}^{\infty}\int_{\mathbf{E}_n\setminus\mathbf{E}_{n-1}}\left(\frac{1-|w|}{|1-\bar{w}z|^2}\right)^p d|\mu|(z)\\ &\preceq \sup_{w\in\mathbf{D}}\sum_{n=1}^{\infty}\frac{|\mu|(\mathbf{E}_n)}{2^{2np}(1-|w|)^p}\\ &\preceq \|\mu\|_{C_p}\sum_{n=1}^{\infty}2^{-np},\end{aligned}$$

that is to say, $\|\mu\|_{C_p} < \infty$.

This lemma provides us a geometric approach to study $\mathcal{Q}_p$.

Theorem 4.1.1. *Let $p \in (0,\infty)$ and $f \in \mathcal{H}$ with*

$$d\mu_{f,p}(z) = |f'(z)|^2(1-|z|^2)^p dm(z), \quad z\in\mathbf{D}.$$

Then $f \in \mathcal{Q}_p$ if and only if $\mu_{f,p}$ is a p-Carleson measure.

Proof. This is a direct by-product of Lemma 4.1.1 and Theorem 1.1.1.

4.2 Relating to Mean Lipschitz Spaces

For $p \in (0,\infty]$, the Hardy space H^p consists of those functions $f \in \mathcal{H}$ for which

$$\|f\|_{H^p} = \sup_{0<r<1} M_p(f,r) < \infty,$$

where

$$M_p(f,r) = \left(\frac{1}{2\pi}\int_{\mathbf{T}} |f(r\zeta)|^p |d\zeta|\right)^{1/p}, \quad p \in (0,\infty);$$

and

$$M_\infty(f,r) = \max_{\zeta\in\mathbf{T}} |f(r\zeta)|, \quad p = \infty.$$

When $p \in [1,\infty]$ and $\alpha \in (0,1]$, we say that $f \in \mathcal{H}$ belongs to $\Lambda(p,\alpha)$ provided

$$M_p(f',r) \preceq (1-r)^{1-\alpha}, \quad r \in (0,1).$$

Lemma 4.2.1. *Let $p \in [1,\infty], \alpha \in (0,1)$ and let $f(z) = \sum_{k=0}^{\infty} a_k z^{n_k}$ lie in HG. Then $f \in \Lambda(p,\alpha)$ if and only if $\sup_{k\in\mathbf{N}} |a_k| n_k^\alpha < \infty$.*

Proof. Assume that $f \in \Lambda(p,\alpha)$ and $r \in (0,1)$. We have

$$n_k a_k = (2\pi i)^{-1} \int_{|z|=r} f'(z) z^{-n_k} dz.$$

For $p \in [1,\infty)$, Hölder's inequality implies that

$$n_k |a_k| \le r^{1-n_k} \|f'_r\|_{H^p} \preceq r^{1-n_k}(1-r)^{\alpha-1}.$$

For $p = \infty$, we just estimate the integral. Choosing $r = 1 - n_k^{-1}$, we obtain $|a_k| n_k^\alpha \preceq 1$.

Conversely, let $|a_k| n_k^\alpha \preceq 1$. Since the number of the Taylor coefficients a_k when $n_k \in I_n = \{j \in \mathbf{N} : 2^n \le j < 2^{n+1}\}$ is at most $[\log_c 2] + 1$, we get

$$M_\infty(f',r) \preceq r^{-1} \sum_{n=0}^{\infty} \sum_{n_j \in I_n} n_j^{1-\alpha} r^{n_j} \preceq r^{-1}(1-r)^{\alpha-1}.$$

Since $M_p(f',r) \le M_\infty(f',r)$, $f \in \Lambda(p,\alpha)$ and the lemma is proved.

The spaces $\Lambda(p,\alpha)$ are called the mean Lipschitz spaces and discussed in [35] where it is proved that the spaces $\Lambda(p,1/p)$ increase with p and are all contained in BMOA. This inclusion suggests a comparison with the $\mathcal{Q}_p$-spaces.

Theorem 4.2.1. *Let $p \in (2,\infty)$ and $q = 1 - 2/p$. Then*
(i) $\Lambda(p,1/p) \subset \bigcap_{\epsilon>0} \mathcal{Q}_{q+\epsilon}$.
(ii) $HG \cap \mathcal{Q}_q \subset \Lambda(p,1/p)$.
(iii) *There exists a function $f \in \mathcal{H}$ satisfying*

$$f \in \bigcap_{q>0} \mathcal{Q}_q \setminus \bigcup_{p<\infty} \Lambda(p,1/p).$$

Proof. (i) We suppose first that $f \in \Lambda(p, 1/p)$ so that $M_p(f', s) \preceq (1-s)^{1/p-1}$, $s \in (0,1)$. Then, for the Carleson box $S(I) = \{z \in \mathbf{D} : 1-h \le |z| < 1, |\theta - \arg z| \le h/2\}$; $h = |I|$, we get by Hölder's inequality and the assumptions $p > 2$ and $q = 1 - 2/p$ that

$$\begin{aligned}
\mu_{f,q}(S(I)) &= \int_{1-h}^{1} \left(\int_{\theta-h/2}^{\theta+h/2} |f'(se^{i\phi})|^2 d\phi \right) (1-s^2)^{q+\epsilon} s ds \\
&\le \int_{1-h}^{1} \left(\int_{\theta-h/2}^{\theta+h/2} |f'(se^{i\phi})|^p d\phi \right)^{2/p} h^{1-2/p}(1-s^2)^{q+\epsilon} ds \\
&\preceq h^{1-2/p} \int_{1-h}^{1} (M_p(f',s))^2 (1-s)^{q+\epsilon} ds \\
&\preceq h^{1-2/p} \int_{1-h}^{1} (1-s)^{2(1/p-1)} (1-s)^{q+\epsilon} ds \approx |I|^{q+\epsilon}.
\end{aligned}$$

By Theorem 4.1.1, $f \in \mathcal{Q}_{q+\epsilon}$ for all $\epsilon > 0$ and thus the inclusion is proved.

In order to prove the strict inclusion, we consider a function $f(z) = \sum_{k=0}^{\infty} a_k z^{n_k}$ where $a_k = k^{1/2} 2^{-k/p}$ and $n_k = 2^k$. Then $|a_k| n_k^{1/p} = n^{1/2}$, and by Lemma 4.2.1, $f \notin \Lambda(p, 1/p)$. On the other hand,

$$\sum_{n=0}^{\infty} 2^{n(1-(q+\epsilon))} \Big(\sum_{n_k \in I_n} |a_k|^2 \Big) = \sum_{n=0}^{\infty} n 2^{-n\epsilon} < \infty,$$

and, by Theorem 1.2.1 (i), $f \in \mathcal{Q}_{q+\epsilon}$ for all $\epsilon \in (0, 2/p]$.

(ii) Suppose that $f(z) = \sum_{k=0}^{\infty} a_k z^{n_k}$ belongs to $HG \cap \mathcal{Q}_{1-2/p}$. By Lemma 4.2.1, it suffices to show $|a_k|^2 n_k^{2/p} \preceq 1$. But this is obvious since the Taylor coefficients a_k of $f \in HG \cap \mathcal{Q}_{1-2/p}$ satisfy (1.4).

To verify the strict inclusion, we construct a Hadamard gap series

$$f_p(z) = \sum_{k=0}^{\infty} a_k z^{n_k} = \sum_{n=0}^{\infty} 2^{-n/p} z^{2^n}.$$

Since $|a_k| n_k^{1/p} = 1$, it follows from Lemma 4.2.1 that $f_p \in \Lambda(p, 1/p)$. On the other hand,

$$\sum_{k=0}^{\infty} n_k^{1-(1-2/p)} \Big(\sum_{n_k \in I_n} |a_k|^2 \Big) = \sum_{n=0}^{\infty} (2^n)^{2/p} 2^{-2n/p} = \infty.$$

By Theorem 1.2.1 (i), $f \notin \mathcal{Q}_{1-2/p}$.

(iii) Hereafter, we use $\|\cdot\|_p$, $p \in (0, \infty)$, to represent the usual L^p-norm. Suppose that we can select a function $f(z) = \sum_{n=0}^{\infty} a_n z^n$ satisfying two conditions below:

(a) $\|\Delta_n f\|_2 \le 2^{-n}$, where $(\Delta_n f)(\zeta) = \sum_{k \in I_n} a_k \zeta^k$ for $n \in \mathbf{N}$ and $\zeta \in \mathbf{T}$;

(b) there exists $n = n(p, m)$, for $p = 3, 4, \cdots$ and $m \in \mathbf{N}$, such that

$$\|\Delta_n f\|_p \geq m 2^{-n/p}.$$

Then for $q > 0$,

$$\begin{aligned}\sum_{n=0}^{\infty} 2^{n(1-q)} \Big(\sum_{k\in I_n} |a_k| \Big)^2 &\leq \sum_{n=0}^{\infty} 2^{n(1-q)} 2^n \sum_{k\in I_n} |a_k|^2 \\ &= \sum_{n=0}^{\infty} 2^{n(1-q)} 2^n \|\Delta_n f\|_2^2 \\ &\leq \sum_{n=0}^{\infty} 2^{-nq} < \infty,\end{aligned}$$

and so $f \in \mathcal{Q}_q$, thanks to the proof of Theorem 1.2.1 (i).

Since the spaces $\Lambda(p, 1/p)$ are monotonically increasing (see also Corollary 2.3 in [35]), it suffices to show that $f \notin \Lambda(p, 1/p)$ for $p = 3, 4, \cdots$. Fix such a p. By (b) there exists $\{n_m\}$ such that $\|\Delta_{n_m} f\|_p \geq m 2^{-m/p}$ for $m \in \mathbf{N}$. Thus, $\sup_n \|\Delta_n f\|_p 2^{n/p} = \infty$ and hence $f \notin \Lambda(p, 1/p)$, by Theorem 3.1 in [35].

The construction. Let $r_1, r_2, \cdots$, be an enumeration of the pairs $\{(p, m) : p = 3, 4, \cdots; m \in \mathbf{N}\}$. We need to find integers n_j: $n_1 < n_2 < \cdots$, and polynomials f_j obeying:

(c) f_j polynomials of degree $\leq 2^{n_j}$;

(d) $\|f_j\|_2 \leq 2^{-n_j}$;

(e) $\|f_j\|_{\pi_1(r_j)} \geq \pi_2(r_j) 2^{-n_j/\pi_1(r_j)}$, where π_j, $j = 1, 2$, are projections on first and second coordinates of the pairs r_j.

Assume that $\{f_j\}$ have been constructed, then define

$$f(z) = \sum_{j=1}^{\infty} f_j(z) z^{2^{n_j}}.$$

It is then easily seen that f satisfies (a) and (b) so we are done once we construct $\{f_j\}$.

Construction of the sequence $\{f_j\}$: Given $n_{j-1}, p = 3, 4, \cdots$, and $m \in \mathbf{N}$, we must find $n_j > n_{j-1}$ and polynomials f_j of degree 2^{n_j} such that

(d)' $\|f_j\|_2 \leq 2^{-n_j}$;

(e)' $\|f_j\|_p \geq m 2^{-n_j/p}$.

But the existence of f_j follows immediately from the density of polynomials in the Hardy space H^p and $H^p \subseteq H^2$, $p > 2$. The proof of the theorem is completed.

4.3 Comparison with Besov Spaces

For $p \in (1, \infty)$, let B_p be the space of all functions $f \in \mathcal{H}$ such that

$$\|f\|_{B_p} = \left(\int_{\mathbf{D}} |f'(z)|^p (1 - |z|^2)^{p-2} dm(z) \right)^{1/p} < \infty.$$

The spaces B_p are the so-called Besov spaces. It is well known that every B_p is conformally invariant according to $\|f \circ \sigma\|_{B_p} = \|f\|_{B_p}$ for all $f \in B_p$ and $\sigma \in Aut(\mathbf{D})$. Of course, it becomes a natural topic to compare $\mathcal{Q}_p$ with B_p.

Lemma 4.3.1. *Let $p \in (1, \infty)$ and let $f(z) = \sum_{k=0}^{\infty} a_k z^{n_k}$ belong to HG. Then $f \in B_p$ if and only if $\sum_{k=0}^{\infty} n_k |a_k|^p < \infty$.*

Proof. The argument is similar to that of Theorem 1.2.1 (i), so we give the key steps of the argument. In fact, if $t_n = \sum_{n_j \in I_n} n_j^2 |a_j|^2$ and $I_n = \{j \in \mathbf{N} : 2^n \leq j < 2^{n+1}\}$, then one has

$$\|f\|_{B_p}^p \approx \sum_{n=0}^{\infty} 2^{-n(p-1)} t_n^{p/2}.$$

Since the number of the Taylor coefficients a_j is at most $[\log_c 2]+1$ when $n_j \in I_n$,

$$t_n^{p/2} \approx 2^{pn} \sum_{n_j \in I_n} |a_j|^p.$$

The above two estimates lead to

$$\|f\|_{B_p}^p \approx \sum_{k=0}^{\infty} n_k |a_k|^p.$$

Theorem 4.3.1. *Let $p \in [1, \infty)$. Then $B_{2p} \subset \bigcap_{1-1/p<q<1} \mathcal{Q}_q$.*

Proof. When $p = 1$, the result follows from Corollary 1.2.1. Accordingly, we only consider $p > 1$ and $1 - 1/p < q < 1$. If $f \in B_{2p}$, then, for the Carleson box $S(I) = \{z \in \mathbf{D} : 1 - h \leq |z| < 1,\ |\theta - \arg z| \leq h/2\}$; $h = |I|$, we get by Hölder's inequality that

$$\begin{aligned}
\mu_{f,q}(S(I)) &= \int_{1-h}^{1} \left(\int_{\theta-h/2}^{\theta+h/2} |f'(re^{i\phi})|^2 d\phi \right) (1 - r^2)^q r dr \\
&\preceq \left(\int_{1-h}^{1} \int_{\theta-h/2}^{\theta+h/2} |f'(re^{i\phi})|^{2p} (1 - r^2)^{2p-2} r d\phi dr \right)^{1/p} h^q \\
&\preceq \|f\|_{B_{2p}}^2 |I|^q
\end{aligned}$$

By Theorem 4.1.1, $f \in \mathcal{Q}_q$ for all $q \in (1 - 1/p, 1)$ and thus the inclusion is proved.

To prove the strict inclusion, we choose a function $f(z) = \sum_{k=0}^{\infty} a_k z^{n_k}$ where $a_k = 2^{-k/(2p)}$ and $n_k = 2^k$. Then $\sum_{k=1}^{\infty} |a_k|^{2p} n_k = \infty$, and by Lemma 4.3.1, $f \notin B_{2p}$. Nevertheless,

$$\sum_{n=0}^{\infty} 2^{n(1-q)} \sum_{n_k \in I_n} |a_k|^2 = \sum_{n=0}^{\infty} 2^{n(1-q-1/p)} < \infty,$$

and, by Theorem 1.2.1 (i), $f \in \mathcal{Q}_q$ for all $q \in (1-1/p, 1)$.

4.4 Mean Growth

The discussions carried out in last two sections suggest a consideration of the mean growth of the derivatives of functions in $\mathcal{Q}_p$.

Theorem 4.4.1. *Let $p \in (0,1]$ and $f \in \mathcal{H}$. If*

$$\int_0^1 (1-r)^p (M_\infty(f',r))^2 dr < \infty,$$

then $f \in \mathcal{Q}_p$. Furthermore, the exponent p cannot be increased, i.e., given $\epsilon > 0$ there exists an $f \in \mathcal{H}$ such that

$$\int_0^1 (1-r)^{p+\epsilon} \big(M_\infty(f',r)\big)^2 dr < \infty,$$

but $f \notin \mathcal{Q}_p$.

Proof. Let $f \in \mathcal{H}$ satisfy

$$\int_0^1 (1-r)^p \big(M_\infty(f',r)\big)^2 dr < \infty.$$

For the Carleson box $S(I) = \{z \in \mathbf{D} : 1-h \le |z| < 1,\ |\theta - \arg z| \le h/2\}$; $h = |I|$, we have

$$\int_{\theta-h/2}^{\theta+h/2} |f'(re^{i\phi})|^2 d\phi \preceq |I| \big(M_\infty(f',r)\big)^2.$$

Hence,

$$\int_{1-h}^1 \left(\int_{\theta-h/2}^{\theta+h/2} |f'(re^{i\phi})|^2 d\phi \right) \frac{r dr}{(1-r^2)^{-p}} \preceq |I|^p \int_0^1 \frac{(M_\infty(f',r))^2}{(1-r)^{-p}} dr.$$

By Theorem 4.1.1, $f \in \mathcal{Q}_p$.

In order to prove that the above exponent p cannot be increased, we take

$$f(z) = \sum_{k=0}^{\infty} 2^{k(p-1)/2} z^{2^k}.$$

Then Theorem 1.2.1 (i) shows $f \notin \mathcal{Q}_p$. Now, it is easy to see that

$$M_\infty(f',r) \preceq (1-r)^{-(p+1)/2}, \quad 0<r<1.$$

Thus, if $\epsilon>0$ then

$$\int_0^1 (1-r)^{p+\epsilon}\big(M_\infty(f',r)\big)^2 dr \preceq \int_0^1 \frac{dr}{(1-r)^{1-\epsilon}} < \infty.$$

This concludes the proof.

From Theorem 4.1.1 it turns out that for $0<p\le 1$,

$$f\in\mathcal{Q}_p \Longrightarrow \int_0^1 (1-r)^p\big(M_2(f',r)\big)^2 dr < \infty.$$

In fact, we can obtain a better result.

Theorem 4.4.2. *Let $p\in(0,1]$, $q\in(0,2]$ and $f\in\mathcal{H}$. If $f\in Q_p$ then*

$$\int_0^1 (1-r)^p\big(M_q(f',r)\big)^2 dr < \infty.$$

Moreover this mean growth is sharp in the following sense: let ϕ be a nonnegative nondecreasing function defined on $(0,1)$ such that

$$\int_0^1 (1-r)^p\phi^2(r)dr < \infty, \tag{4.1}$$

then there exists a function $f\in\mathcal{Q}_p$ such that

$$M_q(f',r)\ge\phi(r), \quad 0<r<1. \tag{4.2}$$

Proof. Since $M_q(f',r)$ is a nondecreasing function of q, it suffices to show the "sharp" part of the theorem. Let p and ϕ be as above. We start with considering the case $q=2$. Set $r_k=1-2^{-k}, k\in\mathbf{N}$. Since ϕ is nondecreasing, we apply (4.1) to get

$$\begin{aligned}\int_0^1 (1-r)^p\phi^2(r)dr &\ge \sum_{k=1}^\infty \int_{r_k}^{r_{k+1}} (1-r)^p\phi^2(r)dr\\ &\ge \sum_{k=1}^\infty (r_{k+1}-r_k)\,(1-r_{k+1})^p\phi^2(r_k)\\ &\approx \sum_{k=1}^\infty 2^{-k(1+p)}\phi^2(r_k).\end{aligned}$$

Thus Theorem 1.2.1 (i) shows that

$$f(z)=\phi(r_1)z+e^4\sum_{k=1}^\infty 2^{-k}\phi(r_k)z^{2^k}$$

is a member of $\mathcal{Q}_p$. However,

$$\left(M_2(f',r)\right)^2 \geq \phi^2(r_1) + e^8 \sum_{k=1}^{\infty} \phi^2(r_k) r^{2^{k+1}}, \quad r \in (0,1). \tag{4.3}$$

This estimate, together with the fact that ϕ is nondecreasing, implies

$$\left(M_2(f',r)\right)^2 \geq \phi^2(r_1) \geq \phi^2(r), \quad 0 < r \leq r_1. \tag{4.4}$$

Also, using the elementary inequality $\left(1-n^{-1}\right)^n \geq e^{-2}$ which is valid for all integers $n \geq 2$, and keeping in mind that ϕ is nondecreasing, we find that (4.3) implies that if $j \geq 1$ and $r_j \leq r \leq r_{j+1}$ then

$$\left(M_2(f',r)\right)^2 \geq e^8 \phi^2(r_{j+1}) r^{2^{j+2}} \geq e^8 \phi^2(r) r_j^{2^{j+2}} \geq \phi^2(r).$$

This, together with (4.4), gives

$$\left(M_2(f',r)\right)^2 \geq \phi^2(r), \quad 0 < r < 1,$$

and hence the proof is done for $q = 2$.

Take now q, $0 < q < 2$. Let f be the function constructed in the previous case. Since f is given by a power series with Hadamard gaps, Theorem 8.20 in [146, p. 215] shows

$$M_q(f',r) \succeq M_2(f',r) \succeq \phi(r), \quad 0 < r < 1.$$

Therefore, the proof is complete.

Notes

4.1 When $p = 1$, Lemma 4.1.1 and Theorem 4.1.1 (taken from Aulaskari-Stegenga-Xiao [24]) are well known and are contained in works of Fefferman, Garcia and Pommerenke, see Baernstein [30] for an exposition on these works. For $p = 2$, see also Xiao [133]. Lemma 4.1.1 can be further applied to study the meromorphic Q classes; see e.g. Aulaskari-Wulan-Zhao [26], Essén-Wulan [61] and Wulan [131]. Theorem 4.1.1 has been extended to the higher dimensions, see Andersson-Carlsson [9] and Yang [141] [142].

Classically, a nonnegative Borel measure μ on $\mathbf{D}$ is a Carleson measure for $\mathcal{D}_p$, $p \in (0,\infty)$, provided

$$\int_{\mathbf{D}} |f|^2 d\mu \preceq \|f\|^2_{\mathcal{D}_p}, \quad f \in \mathcal{D}_p.$$

For $p = 1$, Carleson characterized these measures and applied them to solve the corona theorem [38]. These measures were also important in Fefferman-Stein's duality for H^1 [64]. It is known that for $p \geq 1$, μ is a Carleson measure for $\mathcal{D}_p$

if and only if $\|\mu\|_{C_p} < \infty$; and, if $p \in (0,1)$, then the condition $\|\mu\|_{C_p} < \infty$ is necessary, but not sufficient. In fact, such a Carleson measure is characterized in terms of the Bessel capacity. When $p = 0$, the classical logarithmic capacity must be used. See Stegenga [117] for details. On the other hand, the Carleson measures for $\mathcal{D}_p$ may be described by single Carleson box; see Kerman-Sawyer [84] and Arcozzi-Sawyer-Rochberg [7]. Here, it is also worth mentioning that Ahern and Jevtić [3] used the strong Hausdorff capacity estimates of Adams [1] to obtain that a nonnegative Borel measure μ on $\mathbf{D}$ is a p-Carleson measure if and only if

$$\int_{\mathbf{D}} |f| d\mu \preceq |f(0)| + \|f\|_{B^1_{1-p}}, \quad f \in B^1_{1-p},$$

where a function $f \in \mathcal{H}$ is called to belong to the Besov space B^1_{1-p}, $p \in (0,1)$, provided

$$\|f\|_{B^1_{1-p}} = \int_{\mathbf{D}} |f'(z)|(1-|z|^2)^{p-1} dm(z) < \infty.$$

4.2 Lemma 4.2.1 should be compared with Theorem 1.2.1. And Theorem 4.2.1 improves actually the result that $\Lambda(p, 1/p)$ increases with p and is contained in BMOA; see [35]. For a further discussion, consult Essén-Xiao [62].

4.3 An elementary argument for Theorem 4.3.1 can be found in Aulaskari-Csordas [15]. It is clear that $\mathcal{Q}_p$ and B_p behave similarly in the sense of the limit space: $\lim_{p\to\infty} B_p = \mathcal{B}$ and $\mathcal{Q}_p = \mathcal{B}$ for $p > 1$. Nevertheless, B_p is much smaller than $\mathcal{Q}_p$.

4.4 Theorems 4.4.1 and 4.4.2 come from Aulaskari-Girela-Wulan [18], but also have a close relation with the main results in the paper [71] by Girela-Marquez.

5. Inner-Outer Structure

Based on the classical factorization of the Hardy space H^p, this chapter focuses on: characterizing the inner and outer functions in $\mathcal{Q}_p$; giving the canonical factorization of $\mathcal{Q}_p$; and representing each $\mathcal{Q}_p$-function as the quotient of two functions in $H^\infty \cap \mathcal{Q}_p$.

5.1 Singular Factors

Since $H^\infty \subset$ BMOA, it is a natural question to ask about: is H^∞ a subset of $\mathcal{Q}_p$ for $0 < p < 1$? Unfortunately, this question has a negative answer. To this end, we give some examples.

Example 5.1.1. There are functions in $H^\infty \setminus (\cup_{0<p<1} \mathcal{Q}_p)$.

Proof. For $q \in (1, \infty)$, we consider

$$f_1(z) = \sum_{k=2}^{\infty} (k \log^q k)^{-1} z^{2^k}; \quad f_2(z) = \sum_{k=1}^{\infty} k^{-q} z^{2^k}.$$

For $|z| \leq 1$, both these power series are absolutely convergent and it is clear that f_1 and f_2 are in H^∞. Furthermore, if $a_j = (j \log^q j)^{-1}$, we see that for $p \in (0,1)$,

$$\sum_{k=2}^{\infty} 2^{k(1-p)} \sum_{2^k \leq 2^j < 2^{k+1}} |a_j|^2 = \sum_{k=2}^{\infty} 2^{k(1-p)} a_k^2 = \infty.$$

It follows from Theorem 1.2.1 (i) that $f_1 \notin \bigcup_{0<p<1} \mathcal{Q}_p$. The same kind of argument will show $f_2 \notin \bigcup_{0<p<1} \mathcal{Q}_p$.

Example 5.1.1 suggests us to work with inner functions of H^∞. Recall that a function $B \in H^\infty$ is called an inner function provided $\|B\|_{H^\infty} \leq 1$ and its radial limit $\lim_{r \to 1} |B(r\zeta)| = 1$ for almost all $\zeta \in \mathbf{T}$. It is well known that any inner function lies in BMOA, and yet, this is no longer true in case of $\mathcal{Q}_p$, $p \in (0,1)$. In order to see this, we need an estimation exchanging the derivative and the difference quotient.

Theorem 5.1.1. *Let $p, \delta \in (0,1)$, $\zeta \in \mathbf{T}$ and B be an inner function. Then*

$$\int_\delta^1 |B'(r\zeta)|^2(1-r^2)^p dr \approx \int_\delta^1 (1-|B(r\zeta)|^2)^2(1-r^2)^{p-2}dr. \tag{5.1}$$

Consequently, $B \in \mathcal{Q}_p$ if and only if $(1-|B(z)|^2)^2(1-|z|^2)^{p-2}dm(z)$ is a p-Carleson measure.

Proof. Since B is inner, one always has $(1-|z|^2)|B'(z)| \le 1-|B(z)|^2$ and it suffices to prove the left-hand side inequality of (5.1), i.e.,

$$\int_\delta^1 |B'(r\zeta)|^2(1-r^2)^p dr \preceq \int_\delta^1 (1-|B(r\zeta)|^2)^2(1-r^2)^{p-2}dr.$$

Note that for almost all $\zeta \in \mathbf{T}$,

$$1-|B(r\zeta)| \le |B(\zeta)-B(r\zeta)| = \left|\int_0^1 B'(((1-t)r+t)\zeta)(1-r)dt\right|.$$

Applying Minkowski's inequality, we obtain

$$\begin{aligned}\left(\int_\delta^1 \cdots\right)^{1/2} &= \left(\int_\delta^1 (1-|B(r\zeta)|^2)^2(1-r^2)^{p-2}dr\right)^{1/2}\\ &\le 2\int_0^1 \left(\int_\delta^1 |B'((t+(1-t)r)\zeta)|^2(1-r)^p dr\right)^{1/2} dt\\ &= 2\int_0^1 \left(\int_{t+(1-t)\delta}^1 |B'(s\zeta)|^2\left(\frac{1-s}{1-t}\right)^p \frac{ds}{1-t}\right)^{1/2} dt\\ &\le 2\int_0^1 \left(\int_\delta^1 |B'(s\zeta)|^2(1-s)^p ds\right)^{1/2} \frac{dt}{(1-t)^{(p+1)/2}}\\ &= \frac{4}{1-p}\left(\int_\delta^1 |B'(s\zeta)|^2(1-s)^p ds\right)^{1/2}.\end{aligned}$$

The proof is complete.

Corollary 5.1.1. *Let $p \in (0,1)$, $n \in \mathbf{N}$ and $B = \prod_{j=1}^n B_j$, where each B_j is an inner function. Then $B \in \mathcal{Q}_p$ if and only if $B_j \in \mathcal{Q}_p$ for $j = 1, 2, \dots n$.*

Proof. If $B_j \in \mathcal{Q}_p$ for $j = 1, 2, \dots n$, we use induction to show $B \in \mathcal{Q}_p$. On the other hand, let $B \in \mathcal{Q}_p$. Since we know that $|B_j| \ge |B|$ for all j and that for the Carleson box $S(I)$,

$$\int_{S(I)} (1-|B_j(z)|^2)^2(1-|z|^2)^{p-2}dm(z) \le \int_{S(I)} (1-|B(z)|^2)^2(1-|z|^2)^{p-2}dm(z).$$

Theorem 5.1.1, together with the assumption $B \in \mathcal{Q}_p$, proves $B_j \in \mathcal{Q}_p$, $j = 1, 2, \dots n$.

The forthcoming examples show that all singular inner functions are outside all $\mathcal{Q}_p$, $p \in (0,1)$. Recall that a singular inner function is a function of the form

$$S_s(z) = \exp\Big(\int_{\mathbf{T}} \frac{z+\zeta}{z-\zeta} d\nu_s(\zeta)\Big),$$

where the measure ν_s is nonnegative and singular to $|d\zeta|$, and the index s means "singular".

Example 5.1.2. Let $p \in (0,1)$. Then $S_s \notin \mathcal{Q}_p$.

Proof. First of all, we deal with the simplest case where ν_s is atomic. More explicitly, for $\gamma \in (0,1)$ and $\eta \in \mathbf{T}$ let

$$S_{\gamma,\eta}(z) = \exp\Big(\gamma \frac{z+\eta}{z-\eta}\Big).$$

Then $S_{\gamma,\eta} \notin \mathcal{Q}_p$. In fact, if otherwise, $S_{\gamma,\eta} \in \mathcal{Q}_p$, then for the Carleson box $S(I)$, we get by Theorems 4.1.1 and 5.1.1 that

$$\begin{aligned}
\|S_{\gamma,\eta}\|^2_{\mathcal{Q}_p}|I|^p &\succeq \int_{S(I)} \big(1-|S_{\gamma,\eta}(z)|\big)^2 (1-|z|^2)^{p-2} dm(z)\\
&\geq \int_{S(I)} \Big(1-\exp\big(-\frac{\gamma(1-|z|^2)}{4|I|^2}\big)\Big)^2 (1-|z|^2)^{p-2} dm(z)\\
&\succeq \int_{S(I)} \Big(1-\exp\big(-\frac{\gamma(1-|z|)}{4|I|^2}\big)\Big)^2 (1-|z|)^{p-2} dm(z)\\
&\succeq |I|^{2p-1}\gamma^{1-p} \int_0^{\gamma/(4|I|)} (1-e^{-s})^2 s^{p-2} ds.
\end{aligned}$$

Consequently,

$$\int_0^{\gamma/(4|I|)} (1-e^{-s})^2 s^{p-2} ds \preceq |I|^{1-p}\gamma^{p-1}$$

which is impossible as $|I| \to 0$. Hence $S_{\gamma,\eta} \notin \mathcal{Q}_p$.

Next, let us prove $S_s \notin \mathcal{Q}_p$. If S_s contains a factor $S_{\gamma,\zeta}$, from Corollary 5.1.1 and the above simple case it follows that $S_s \notin \mathcal{Q}_p$. It remains to consider the case when ν_s is nonatomic. Let

$$\omega(\delta) = \sup\{\nu_s(I) : I \text{ subarc of } \mathbf{T}, |I| = \delta\}.$$

We know that $\omega(\delta)$ is continuous and that $\lim_{\delta\to 0}\omega(\delta)/\delta = \infty$ (see Theorem 8.11 in [108]). If $S_s \in \mathcal{Q}_p$, then Theorem 5.1.1 holds with B replaced by S_s. Consequently, for the Carleson box $S(I)$ with $|I| = \delta < 1/2$ one has

$$\begin{aligned}\|S_s\|^2_{\mathcal{Q}_p}|I|^p &\succeq \int_{S(I)}(1-|S_s(z)|^2)^2(1-|z|^2)^{p-2}dm(z)\\ &\succeq \int_{S(I)}\left(1-\exp\left(-2\int_{\mathbf{T}}\frac{1-|z|^2}{|1-\bar{\zeta}z|^2}d\nu_s(\zeta)\right)\right)^2(1-|z|^2)^{p-2}dm(z)\\ &\succeq \int_{S(I)}\left(1-\exp\left(-\frac{\omega(\delta)(1-|z|^2)}{4\delta^2}\right)\right)^2(1-|z|^2)^{p-2}dm(z)\\ &\succeq |I|\int_{1-\delta}^{1}\left(1-\exp\left(-\frac{\omega(\delta)(1-r)}{4\delta^2}\right)\right)^2(1-r)^{p-2}dr\\ &\succeq |I|^p\left(\frac{\delta}{\omega(\delta)}\right)^{p-1}\int_0^{\omega(\delta)/(4\delta)}(1-e^{-t})^2t^{p-2}dt,\end{aligned}$$

which deduces immediately that

$$\int_0^{\omega(\delta)/4\delta}(1-e^{-t})^2t^{p-2}dt \preceq \|S_s\|^2_{\mathcal{Q}_p}\left(\frac{\delta}{\omega(\delta)}\right)^{1-p}\to 0$$

as $\delta\to 0$. This is a contradiction, and so our assumption that $S_s\in\mathcal{Q}_p$ is wrong. The proof is complete.

5.2 Blaschke Products

Corresponding to those singular inner functions are the Blaschke products. By definition, a Blaschke product on $\mathbf{D}$ is a function of the form

$$B(\{z_n\},z)=\prod_{n\in\mathbf{N}}\frac{|z_n|}{z_n}\frac{z_n-z}{1-\bar{z}_nz},$$

where $\{z_n\}\subset\mathbf{D}$ is a sequence satisfying the condition $\sum_{n=1}^{\infty}(1-|z_n|^2)<\infty$. Note that if $z_n=0$ then $|z_n|/z_n$ is replaced by 1. Of course, every Blaschke product is an inner function, but not all Blaschke products are in $\mathcal{Q}_p$ because [107, Theorem 1] tells us that there exists a Blaschke product B satisfying $\int_{\mathbf{D}}|B'(z)|dm(z)=\infty$. The existence of such a Blaschke product B leads to that

$$\infty=\int_{\mathbf{D}}|B'(z)|dm(z)\le F_p(B,0)\left(\int_{\mathbf{D}}(1-|z|^2)^{-p}dm(z)\right)^{1/2}$$

holds for $p\in(0,1)$. This implies $B\notin\mathcal{Q}_p$. A careful analysis reveals that every Blaschke product in $\mathcal{Q}_p$ is closely related to its zeros. The precise relation can be presented by a full description of all inner functions in $\mathcal{Q}_p$.

Theorem 5.2.1. *Let $p\in(0,1)$ and B be inner. Then $B\in\mathcal{Q}_p$ if and only if B is a Blaschke product whose zero set $\{z_n\}$ is such that $\mu_{\{z_n\},p}$ is a p-Carleson measure, where*

$$d\mu_{\{z_n\},p} = \sum_{n=1}^{\infty}(1-|z_n|^2)^p \delta_{z_n}$$

and δ_ζ denotes a Dirac measure at ζ.

Proof. Necessity: assuming that $B \in \mathcal{Q}_p$ which, by Theorems 4.1.1 and 5.1.1, leads to

$$\int_{S(I)} (1-|B(z)|^2)^2(1-|z|^2)^{p-2}dm(z) \preceq \|B\|^2_{\mathcal{Q}_p}|I|^p \tag{5.2}$$

for the Carleson box $S(I)$, we claim that B must be a Blaschke product. In fact, we will prove that if B is not a Blaschke product, then it follows that $B \notin \mathcal{Q}_p$. By the classical factorization theorem (cf. [66, p.74, Theorem 5.5]), every inner function B can be represented as a product of a complex number $\eta \in \mathbf{T}$, a Blaschke product and a singular inner function generated by a singular measure ν_s on $\mathbf{T}$:

$$S_s(z) = \exp\Big(\int_{\mathbf{T}} \frac{z+\zeta}{z-\zeta} d\nu_s(\zeta)\Big).$$

By Example 5.1.2 we know that S_s is not a member of $\mathcal{Q}_p$. Furthermore, it turns out from Corollary 5.1.1 that each inner function $B \in \mathcal{Q}_p$ contains its Blaschke product only. It remains to prove that $\mu_{\{z_n\},p}$ is a p-Carleson measure. In the following argument, we assume without loss of generality that $|I| \le 1/2$. To shorten our formulas, we introduce

$$T(\{z_n\},z) = \sum_{n=1}^{\infty}(1-|\sigma_{z_n}(z)|^2)$$

$$R(\{z_n\},I) = \sum_{z_n\in S(I)}(1-|z_n|^2).$$

Since

$$\log|B(z)|^2 = 2\sum_{n=1}^{\infty}\log\Big(1-\frac{(1-|z_n|^2)(1-|z|^2)}{|1-\overline{z}_n z|^2}\Big) \le -2T(\{z_n\},z),$$

we have also

$$1-|B(z)|^2 \ge 1-\exp\{-2T(\{z_n\},z)\}. \tag{5.3}$$

Combining (5.2) and (5.3), we see

$$\begin{aligned}\|B\|^2_{\mathcal{Q}_p}|I|^p &= \int_{S(I)}\Big(1-\exp(-2T(\{z_n\},z)\Big)^2(1-|z|^2)^{p-2}dm(z)\\ &\ge \int_{S(I)}\Big(1-\exp\Big(-\frac{(1-|z|^2)R(\{z_n\},I)}{4|I|^2}\Big)\Big)^2(1-|z|^2)^{p-2}dm(z)\\ &\succeq |I|\int_0^{|I|}\Big(1-\exp\Big(-\frac{rR(\{z_n\},I)}{4|I|^2}\Big)\Big)^2 r^{p-2}dr\end{aligned}$$

$$\succeq |I| \int_0^{R(\{z_n\},I)/(4|I|)} (1-e^{-s})^2 s^{p-2} \Big(\frac{|I|^2}{R(\{z_n\},I)}\Big)^{p-1} ds$$
$$\succeq \Big(\frac{R(\{z_n\},I)}{|I|}\Big)^{1-p} |I|^p \int_0^{R(\{z_n\},I)/(4|I|)} (1-e^{-s})^2 s^{p-2} ds.$$

This implies

$$\Big(\frac{R(\{z_n\},I)}{|I|}\Big)^{1-p} \int_0^{R(\{z_n\},I)/(4|I|)} (1-e^{-s})^2 s^{p-2} ds \preceq \|B\|_{\mathcal{Q}_p}^2,$$

so that $\sum_{n=1}^{\infty}(1-|z_n|^2)\delta_{z_n}$ is a 1-Carleson measure, namely,

$$M = \sup_{z\in\mathbf{D}} T(\{z_n\}, z) < \infty,$$

owing to Lemma 4.1.1.

Because

$$\frac{1-e^{-2t}}{2t} \geq \frac{1-e^{-2M}}{2M} = M_1, \quad 0 \leq t \leq M,$$

it follows from (5.3) that $1-|B(z)|^2 \geq 2M_1 T(\{z_n\}, z)$, so that

$$1-|B\circ\sigma_w(z)|^2 \succeq M_1 T(\{\sigma_w(z_n)\}, z).$$

Using the assumption $B \in \mathcal{Q}_p$, Theorems 1.1.1 and 5.1.1, and Lemma 1.4.1, we get that for $w \in \mathbf{D}$,

$$\begin{aligned}(F_p(B,w))^2 &= \int_{\mathbf{D}} |(B\circ\sigma_w)'(z)|^2(1-|z|^2)^p dm(z)\\ &\succeq \int_{\mathbf{D}} (1-|B\circ\sigma_w(z)|^2)^2(1-|z|^2)^{p-2} dm(z)\\ &\succeq \int_{\mathbf{D}} \big(T(\{\sigma_w(z_n)\},z)\big)^2(1-|z|^2)^{p-2} dm(z)\\ &\succeq \sum_{n=1}^{\infty}(1-|\sigma_w(z_n)|^2)^2 \int_{\mathbf{D}} \frac{(1-|z|^2)^p}{|1-z\overline{\sigma_w(z_n)}|^4} dm(z)\\ &\succeq \sum_{n=1}^{\infty}(1-|\sigma_w(z_n)|^2)^p.\end{aligned}$$

These inequalities imply $\|\mu_{\{z_n\},p}\|_{\mathcal{C}_p} < \infty$. In other words, $\mu_{\{z_n\},p}$ is a p-Carleson measure, due to Lemma 4.1.1.

Sufficiency: if the inner function B is a Blaschke product whose zeros $\{z_n\}$ are such that $\mu_{\{z_n\},p}$ is a p-Carleson measure, then we show that B belongs to $\mathcal{Q}_p$. Writing $B(z) = B(\{z_n\}, z)$, we obtain

$$\left|\frac{B'(z)}{B(z)}\right| = \left|\sum_{n=1}^{\infty} \frac{(1-|z_n|^2)}{(z_n-z)(1-\bar{z}_n z)}\right|$$

and thus

$$|B'(z)| \le \sum_{n=1}^{\infty} \frac{1-|z_n|^2}{|1-\bar{z}_n z|^2}.$$

Applying the conformal invariance of $\|\cdot\|_{\mathcal{B}}$, as well as Lemma 1.4.1, we see that for $w \in \mathbf{D}$,

$$\begin{aligned}(F_p(B,w))^2 &\le \|B\|_{\mathcal{B}}^2 \int_{\mathbf{D}} |(B\circ\sigma_w)'(z)|(1-|z|^2)^{p-1}dm(z)\\ &\le \sum_{n=1}^{\infty}(1-|\sigma_w(z_n)|^2)\int_{\mathbf{D}} \frac{(1-|z|^2)^{p-1}}{|1-z\overline{\sigma_w(z_n)}|^2}dm(z)\\ &\preceq \sum_{n=1}^{\infty}(1-|\sigma_w(z_n)|^2)^p \preceq \|\mu_{\{z_n\},p}\|_{\mathcal{C}_p}.\end{aligned}$$

This, together with Theorem 1.1.1, implies $B \in \mathcal{Q}_p$. The proof is complete.

5.3 Outer Functions

An outer function for H^2 is the function of the form

$$\mathcal{O}_\psi(z) = \eta \exp\left(\int_{\mathbf{T}} \frac{\zeta+z}{\zeta-z}\log\psi(\zeta)\frac{|d\zeta|}{2\pi}\right), \quad \eta \in \mathbf{T},$$

where $\psi > 0$ a.e. on $\mathbf{T}$, $\log\psi \in L^1(\mathbf{T})$ and $\psi \in L^2(\mathbf{T})$.

In what follows, for $z \in \mathbf{D}$ and $\zeta \in \mathbf{T}$ let

$$d\lambda(z) = \frac{dm(z)}{(1-|z|^2)^2} \quad \textit{resp.} \quad d\mu_z(\zeta) = \frac{1-|z|^2}{|\zeta-z|^2}\frac{|d\zeta|}{2\pi}$$

be the hyperbolic measure on $\mathbf{D}$ resp. the Poisson measure on $\mathbf{T}$.

Before giving a description of the outer functions in $\mathcal{Q}_p$, we present a new characterization of $\mathcal{Q}_p$.

Theorem 5.3.1. *Let $p \in (0,1)$ and $f \in H^2$. Then $f \in \mathcal{Q}_p$ if and only if one of the following conditions holds:*

(i)

$$\sup_{w\in\mathbf{D}} \int_{\mathbf{D}} \left(\int_{\mathbf{D}} |f'(u)|^2 g(u,z)dm(u)\right)(1-|\sigma_w(z)|^2)^p d\lambda(z) < \infty.$$

(ii)

$$\sup_{w\in\mathbf{D}} \int_{\mathbf{D}} \left(\int_{\mathbf{T}} |f-f(z)|^2 d\mu_z\right)(1-|\sigma_w(z)|^2)^p d\lambda(z) < \infty.$$

(iii)

$$\sup_{w\in\mathbf{D}} \int_{\mathbf{D}} \left(\int_{\mathbf{T}} |f|^2 d\mu_z - |f(z)|^2\right)(1-|\sigma_w(z)|^2)^p d\lambda(z) < \infty.$$

Proof. Since $\lim_{z\to\mathbf{T}}(1-|z|^2)=0$, we apply Green's theorem [66, p. 236] to get that for $z\in\mathbf{D}$, and $p\in(0,1)$,

$$\begin{aligned}(1-|z|^2)^p &= \frac{2}{\pi}\int_{\mathbf{D}}\Big(\frac{\partial^2}{\partial u\partial\bar{u}}(1-|u|^2)^p\Big)\log|\sigma_z(u)|dm(u)\\ &= \frac{2p}{\pi}\int_{\mathbf{D}}(1-p|u|^2)(1-|u|^2)^p g(u,z)d\lambda(u).\end{aligned}$$

Furthermore, from Fubini's theorem, Hardy-Littlewood's identity (cf. [66, p. 238, (3.3)]), it follows that

$$\begin{aligned}\frac{\pi(F_p(f,0))^2}{2p} &= \int_{\mathbf{D}}\Big(\int_{\mathbf{D}}|f'(z)|^2 g(u,z)dm(z)\Big)(1-p|u|^2)(1-|u|^2)^p d\lambda(u)\\ &\approx \int_{\mathbf{D}}(1-|u|^2)^p\Big(\int_{\mathbf{T}}|f(\zeta)-f(u)|^2 d\mu_z(\zeta)\Big)d\lambda(u)\\ &= \int_{\mathbf{D}}(1-|u|^2)^p\Big(\int_{\mathbf{T}}|f(\zeta)|^2 d\mu_u(\zeta)-|f(u)|^2\Big)d\lambda(u).\end{aligned}$$

Since $d\mu_z$ and $d\lambda$ are conformally invariant, when $\zeta\in\mathbf{T}$, $w,z\in\mathbf{D}$ one has

$$d\mu_z(\zeta)=d\mu_{\sigma_w(z)}(\sigma_w(\zeta)) \quad \textit{and} \quad d\lambda(z)=d\lambda(\sigma_w(z)).$$

Theorem 1.1.1, with the help of some elementary calculations, implies the desired equivalence.

Concerning the outer functions in $\mathcal{Q}_p$, $p\in(0,1)$, we can establish the following result.

Theorem 5.3.2. *Let $p\in(0,1)$ and let $\psi>0$ a.e. on $\mathbf{T}$, $\log\psi\in L^1(\mathbf{T})$ and $\psi\in L^2(\mathbf{T})$. Then $\mathcal{O}_\psi\in\mathcal{Q}_p$ if and only if*

$$\sup_{w\in\mathbf{D}}\int_{\mathbf{D}}\Big(\int_{\mathbf{T}}\psi^2 d\mu_z-\exp\Big(\int_{\mathbf{T}}\log\psi^2 d\mu_z\Big)\Big)(1-|\sigma_w(z)|^2)^p d\lambda(z)<\infty.$$

Proof. This follows immediately from both Theorem 5.3.1 (iii) with $f=\mathcal{O}_\psi$ and the fact that $|\mathcal{O}_\psi|=\psi$ a.e. on $\mathbf{T}$ and

$$|\mathcal{O}_\psi(z)|=\exp\Big(\int_{\mathbf{T}}\log\psi d\mu_z\Big),\quad z\in\mathbf{D}.$$

5.4 Canonical Factorization

One of the most essential properties on H^2 is that H^2 has an inner-outer factorization. That is to say, every nonzero H^2-function can be factored in the form $f=B\mathcal{O}$, where B is inner and $\mathcal{O}\in H^2$ is outer. Conversely, such a function $B\mathcal{O}$ belongs to H^2. Then, it is natural to ask: how about $\mathcal{Q}_p$?. Now, Theorems 5.1.2 and 5.3.2 enable us to derive a solution to this question.

Theorem 5.4.1. *Let $p \in (0,1)$ and let $f \in H^2$ with $f \not\equiv 0$. Then $f \in \mathcal{Q}_p$ if and only if $f = B\mathcal{O}$, where B is an inner function and $\mathcal{O} \in \mathcal{Q}_p$ is an outer function for which*

$$\sup_{w\in\mathbf{D}} \int_{\mathbf{D}} |\mathcal{O}(z)|^2(1-|B(z)|^2)(1-|\sigma_w(z)|^2)^p d\lambda(z) < \infty. \tag{5.4}$$

Proof. Because f is a member of H^2 with $f \not\equiv 0$, f must be of the form $B\mathcal{O}$, where B is an inner function and $\mathcal{O}$ is an outer function for H^2. A simple computation produces a formula related to pointwise multiplication below:

$$\int_{\mathbf{T}} |B\mathcal{O}|^2 d\mu_z - |B(z)\mathcal{O}(z)|^2 = \int_{\mathbf{T}} |\mathcal{O}|^2 d\mu_z - |\mathcal{O}(z)|^2 + |\mathcal{O}(z)|^2(1-|B(z)|^2)$$

Thanks to the fact that $|B(z)| \le 1$ for all $z \in \mathbf{D}$, $f \in \mathcal{Q}_p$ is equivalent to that $\mathcal{O}$ lies in $\mathcal{Q}_p$ and $|\mathcal{O}|^2(1-|B|^2)$ meets the requirement of (5.4). The proof is complete.

Theorem 5.4.1 has indeed illustrated that every nonzero $\mathcal{Q}_p$-function f has a unique representation $B\mathcal{O}$, but also the outer factor $\mathcal{O}$ inherits the smoothness of f.

Corollary 5.4.1. *Let $p \in (0,1)$ and let $f \in H^2$ be such that $f/B \in H^2$ for an inner function B. If $f \in \mathcal{Q}_p$ then $f/B \in \mathcal{Q}_p$.*

Proof. Since $f = (f/B)B$, the corollary follows from Theorem 5.4.1.

Theorems 5.3.1 and 5.3.2 will be used to construct the cut-off outer functions in $\mathcal{Q}_p$ and hence to represent every $\mathcal{Q}_p$-function as the ratio of two H^∞-functions in $\mathcal{Q}_p$. To see this, we cite Aleman's lemma as follows.

Lemma 5.4.1. *Let (X,μ) be a probability space and let $\psi \in L^1(\mu)$ with $\psi > 0$ μ-a.e. on X and $\log\psi \in L^1(\mu)$. Let*

$$E(\psi) = \int_X \psi d\mu - \exp\left(\int_X \log\psi d\mu\right).$$

Then

$$\max\{E(\min\{1,\psi\}),\quad E(\max\{1,\psi\})\} \le E_\gamma(\psi). \tag{5.5}$$

Proof. Without loss of generality, assume that $A = \{x \in X : \psi(x) \ge 1\}$ and $\alpha = \mu(A) \in (0,1)$. The inequality $E(\min\{1,\psi\}) \le E_\gamma(\psi)$ is equivalent to

$$\int_A (\psi-1)d\mu \ge \left(\exp\left(\int_{X\setminus A} \log\psi d\mu\right)\right)\left(\exp\left(\int_A \log\psi d\mu\right) - 1\right).$$

Note that the right-hand side of the last inequality is less than or equal to

$$\left(\exp\left(\frac{1}{\alpha}\int_A \log\psi d\mu\right)\right)^{\alpha} - 1 \le \left(\frac{1}{\alpha}\int_A \psi d\mu\right)^{\alpha} - 1$$

due to Jensen's inequality. Now it is easy to show by differentiation that $t-\beta \ge (t/\beta)^\beta - 1$ holds for $t \ge 0$ and $\beta \in (0,1)$. Thus the inequality $E(\min\{1,\psi\}) \le E_\gamma(\psi)$ follows from $t = \int_A \psi d\mu$ and $\beta = \alpha$.

The proof for the other inequality in (5.5) can be given similarly with A replaced by $X \setminus A$ and α by $1-\alpha$.

Theorem 5.4.2. *Let $p \in (0,1)$. Then every function in $\mathcal{Q}_p$ is the quotient of two functions in $H^\infty \cap \mathcal{Q}_p$.*

Proof. The proof relies upon the constructions of two cut-off outer functions attached to a given outer function. To this end, suppose that $f \in \mathcal{Q}_p$ is such that $f \not\equiv 0$ (otherwise there is nothing to argue). Let $B\mathcal{O}$ be the inner-outer factorization of f as in Theorem 5.4.1. In particular,

$$\mathcal{O}(z) = \eta \exp\left(\int_{\mathbf{T}} \frac{\zeta+z}{\zeta-z}\log(|\mathcal{O}(\zeta)|)\frac{|d\zeta|}{2\pi}\right),$$

where $\eta \in \mathbf{T}$ and $z \in \mathbf{D}$. This outer factor $\mathcal{O}$ is equipped with two cut-off outer functions below:

$$\mathcal{O}_+(z) = \sqrt{\eta}\exp\left(\int_{\mathbf{T}} \frac{\zeta+z}{\zeta-z}\log(\max\{|\mathcal{O}(\zeta)|,1\})\frac{|d\zeta|}{2\pi}\right)$$

and

$$\mathcal{O}_-(z) = \sqrt{\eta}\exp\left(\int_{\mathbf{T}} \frac{\zeta+z}{\zeta-z}\log(\min\{|\mathcal{O}(\zeta)|,1\})\frac{|d\zeta|}{2\pi}\right).$$

It is clear that $\mathcal{O}_-$ and $1/\mathcal{O}_+$ lie in H^∞. A key observation is that $\mathcal{O} = \mathcal{O}_+\mathcal{O}_-$, $|\mathcal{O}_-(z)| \le |\mathcal{O}(z)|$ and $|\mathcal{O}_+(z)| \ge 1$ for all $z \in \mathbf{D}$.

For convenience, we put

$$E(\psi, z) = \int_{\mathbf{T}} \psi d\mu_z - \exp\left(\int_{\mathbf{T}} \log\psi d\mu_z\right),$$

where $\psi > 0$ a.e on $\mathbf{T}$ and $\psi, \log\psi \in L^1(T)$.

Since μ_z is a probability measure on $\mathbf{T}$, Lemma 5.4.1 shows

$$E(|\mathcal{O}_+|^2, z) \le E(|\mathcal{O}|^2, z); \quad E(|\mathcal{O}_-|^2, z) \le E(|\mathcal{O}|^2, z).$$

Notice that f can be rewritten as $B\mathcal{O}_+\mathcal{O}_- = (B\mathcal{O}_-)/(1/\mathcal{O}_+)$. Accordingly, it suffices to verify that both $g = B\mathcal{O}_-$ and $h = 1/\mathcal{O}_+$ are members of $\mathcal{Q}_p$. On the one hand, $h \in \mathcal{Q}_p$ is obvious. As a matter of fact, owing to $h' = -\mathcal{O}_+'/\mathcal{O}_+^2$, $|\mathcal{O}_+| \ge 1$ and

$$\begin{aligned}\int_{\mathbf{T}} |\mathcal{O}_+|^2 d\mu_z - |\mathcal{O}_+(z)|^2 &= E(|\mathcal{O}_+|^2, z) \le E(|\mathcal{O}|^2, z) \\ &\le \int_{\mathbf{T}} |\mathcal{O}|^2 d\mu_z - |\mathcal{O}(z)|^2 + |\mathcal{O}(z)|^2(1 - |B(z)|^2) \\ &\le \int_{\mathbf{T}} |f|^2 d\mu_z - |f(z)|^2,\end{aligned}$$

it follows from Theorems 5.3.1 and 5.3.2 that $\mathcal{O}_+ \in \mathcal{Q}_p$ and hence $h \in \mathcal{Q}_p$. On the other hand, $g \in \mathcal{Q}_p$ comes from Theorems 5.3.1 and 5.3.2 as well as the estimates below:

$$\begin{aligned}\int_{\mathbf{T}} |B\mathcal{O}_-|^2 d\mu_z - |B(z)\mathcal{O}_-(z)|^2 &= E(|\mathcal{O}_-|^2, z) + |\mathcal{O}_-(z)|^2(1 - |B(z)|^2) \\ &\le E(|\mathcal{O}|^2, z) + |\mathcal{O}(z)|^2(1 - |B(z)|^2) \\ &= \int_{\mathbf{T}} |\mathcal{O}|^2 d\mu_z - |\mathcal{O}(z)|^2 + |\mathcal{O}(z)|^2(1 - |B(z)|^2) \\ &\le \int_{\mathbf{T}} |f|^2 d\mu_z - |f(z)|^2.\end{aligned}$$

This concludes the proof.

Notes

5.1 The first section of Chapter 5 is from Essén-Xiao [63]. For a result analogous to Theorem 5.1.1, we refer to Verbitskii [125, Lemma 2.2] which was proved by using Hölder's inequality. For an inner function B, the quotient $(1-|B'(z)|^2)/(1-|z|^2)$ has an operator-theoretic explanation. Let $k_w(z) = (1-|w|^2)^{1/2}/(1-\bar{w}z)$ be the normalized reproducing kernel of the Hardy space H^2 with respect to $w \in \mathbf{D}$. If C_B^* denotes the adjoint of the composition operator C_B with the symbol B, then

$$\|C_B^* k_w\|_{H^2}^2 = (1-|w|^2)/(1-|B'(w)|^2)$$

holds for every inner function B. For a proof, see also Shapiro [110, p.43-44].

5.2 Theorem 5.2.1 is one of the main results in Essén-Xiao [63]. As one of its consequences, it was proved by Nicolau-Xiao [95] that any Blaschke product in $\mathcal{Q}_p$ has small mean variation on many subarcs of $\mathbf{T}$. That is to say, if $B \in \mathcal{Q}_p$ is a Blaschke product with $\{z_n\}$ and $\mathbf{L}_r = \{z \in \mathbf{D} : \inf_n |\sigma_z(z_n)| \ge r\}$, $r \in (0,1)$, then

$$\lim_{r\to 1} \sup_{I \subseteq \mathbf{T}} |I|^{-p} \int_{S(I)} |B'(z)|^2 (1-|z|^2)^p 1_{\mathbf{L}_r}(z) dm(z) = 0.$$

Moreover, Resendis and Tovar [104] showed that if $\sum_{n=1}^{\infty}(1-|z_n|^2)^p$ is a p-subordinated, i.e.,

$$\sum_{n=k+1}^{\infty}(1-|z_n|^2)^p \preceq (1-|z_k|^2)^p, \quad k\in\mathbf{N},$$

then $\mu_{\{z_n\},p}$ is a p-Carleson measure. So, the criterion in Theorem 5.2.1 may be checked by using some appropriate requirements only on the distribution of $\{z_n\}$. For more information, see Resendis-Tovar [104], Danikas-Mouratides [46] and Aulaskari-Wulan-Zhao [26]. The characterization of the inner functions in $\mathcal{D}_p$, $p\in[0,1)$ (in terms of the zero distribution of such functions) can be found in Ahern [2] and Carleson [37].

For $f\in\mathcal{H}$ with $f(z)=\sum_{n=0}^{\infty}a_nz^n$; $z\in\mathbf{D}$, the Cesàro operator $\mathcal{C}$ is defined by

$$(\mathcal{C}f)(z)=\sum_{n=0}^{\infty}\Big((n+1)^{-1}\sum_{k=0}^{n}a_n\Big)z^n.$$

Although H^∞ is not a subspace of $\mathcal{Q}_p$, $p\in(0,1)$, the Cesàro operator maps H^∞ into $\mathcal{Q}_p$ thanks to

$$z(\mathcal{C}f)'(z)=\frac{f(z)}{1-z}-\int_0^1\frac{f(tz)}{1-tz}dt.$$

See Essén-Xiao [63, Theorem 5.4], as well as Danikas-Siskakis [48] for BMOA. For more information about the Cesàro operators acting on different holomorphic function spaces, we refer to Siskakis [111] and Benke-Chang [31].

5.3 The third section of Chapter 5 is from Xiao's paper [134] (see also Xiao [138] for meromorphic case). Notice that $f\in$ BMOA if and only if $f=B\mathcal{O}$, where B is an inner function and $\mathcal{O}$ is an outer function in BMOA for which $|\mathcal{O}(z)|^2(1-|B(z)|^2)$ is bounded on $\mathbf{D}$. This result is due to Dyakonov [55].

Theorem 5.3.1 is important since it gives a way to recognize $\mathcal{Q}_p$, $p\in(0,1)$ via BMOA. This theorem can be used to study some isoperimetric inequalities involved in $\mathcal{Q}_p$; see the paper [23] of Aulaskari, Perez-Gonzaléz and Wulan. In particular, if $f\in$BMOA has a hyperbolic image region: $\Omega=f(\mathbf{D})$ for which the Green function is denoted by $g_\Omega(\cdot,\cdot)$, then the condition

$$\sup_{w\in\mathbf{D}}\int_{\mathbf{D}}\Big(\int_\Omega g_\Omega(u,f(z))dm(u)\Big)(1-|\sigma_w(z)|^2)^pd\lambda(z)<\infty$$

implies $f\in\mathcal{Q}_p$, $p\in(0,1)$. Its converse is valid for every universal covering map. In case of BMOA, we refer to Metzger's paper [91] as well as Gotoh's preprint [72]. Regarding the Q classes on Riemann surfaces, we refer the interested reader to Aulaskari-He-Ristioja-Zhao [19], Aulaskari-Chen [14] and references therein.

5.4 With regard to the last section of Chapter 5, we mention that all $\mathcal{D}_p$, $p\in[0,1)$, have analogous results, see Aleman [5] and Dyakonov [54]. Theorem 5.4.2 depends on Lemma 5.4.1. But, the known fact that any BMOA-function equals the ratio of two functions in H^∞ can be also worked out from the corresponding decomposition of the Nevanlinna functions [50]).

6. Pseudo-holomorphic Extension

In this chapter we first give a full boundary value description that f is in $\mathcal{Q}_p$, $p \in (0,1)$, and secondly provide a characterization of $\mathcal{Q}_p$ via the pseudo-holomorphic extension and, as a corollary, we prove that $\mathcal{Q}_p$ has the $\mathcal{K}$-property. The latter means that, for any $\psi \in H^\infty$, the Toeplitz operator $T_{\bar{\psi}}$ maps $\mathcal{Q}_p$ into itself.

6.1 Boundary Value Behavior

A good way to know much more information about $\mathcal{Q}_p$ is to find out how a $\mathcal{Q}_p$-function behaves on $\mathbf{T}$. To understand this view-point, we, from now on, assume that for $\rho > 0$ and an arc $I \subseteq \mathbf{T}$, ρI denotes the subarc of $\mathbf{T}$ with the same center as I and with the arclength $\rho|I|$, but also we need a description of the boundary value functions of elements in $\mathcal{D}_p$.

Lemma 6.1.1. *Let $p \in (0,1)$ and $f \in H^2$. Then $f \in \mathcal{D}_p$ if and only if*

$$\|f\|^2_{\mathcal{D}_p,*} = \int_{\mathbf{T}}\int_{\mathbf{T}} \frac{|f(\zeta) - f(\eta)|^2}{|\zeta - \eta|^{2-p}} |d\zeta||d\eta| < \infty.$$

Proof. Since $f \in H^2$, we may assume $f(z) = \sum_{n=0}^{\infty} a_n z^n$. A simple calculation involving Parseval's formula implies that for each $\zeta \in \mathbf{T}$,

$$\int_{\mathbf{T}} |f(z\bar{\zeta}) - f(z)|^2 |dz| \approx \sum_{n=0}^{\infty} |a_n|^2 |\zeta^n - 1|^2.$$

This estimation leads to

$$\begin{aligned}
\|f\|^2_{\mathcal{D}_p,*} &= \int_{\mathbf{T}}\int_{\mathbf{T}} \frac{|f(z) - f(w)|^2}{|z - w|^{2-p}} |dz||dw| \\
&\approx \int_{\mathbf{T}} \left(\int_{\mathbf{T}} |f(z\bar{\zeta}) - f(z)|^2 |dz| \right) |\zeta - 1|^{p-2} |d\zeta| \\
&\approx \sum_{n=0}^{\infty} |a_n|^2 \int_{\mathbf{T}} |\zeta^n - 1|^2 |\zeta - 1|^{p-2} |d\zeta| \\
&\approx \sum_{n=0}^{\infty} |a_n|^2 n^{1-p} \approx \|f\|^2_{\mathcal{D}_p},
\end{aligned}$$

so that Lemma 6.1.1 follows.

By Lemma 6.1.1 and Theorem 1.1.1 we obtain a characterization of the boundary function of f in $\mathcal{Q}_p$.

Theorem 6.1.1. *Let $p \in (0,1)$ and let $f \in H^2$. Then $f \in \mathcal{Q}_p$ if and only if*

$$\|f\|^2_{\mathcal{Q}_p,*} = \sup_{I \subseteq \mathbf{T}} |I|^{-p} \int_I \int_I \frac{|f(\zeta)-f(\eta)|^2}{|\zeta-\eta|^{2-p}} |d\zeta||d\eta| < \infty,$$

where the supremum is taken over all arcs $I \subseteq \mathbf{T}$.

Proof. By a change of variables: $z = \sigma_w(u)$, $w \in \mathbf{D}$, we can easily establish

$$\frac{\|f\circ\sigma_w\|^2_{\mathcal{D}_p,*}}{(2\pi)^p} = \int_{\mathbf{T}} \int_{\mathbf{T}} \frac{|f(u)-f(v)|^2}{|u-v|^{2-p}} \left(P_w(u)P_w(v)\right)^{\frac{p}{2}} |du||dv|, \tag{6.1}$$

where

$$2\pi P_w(u) = \frac{1-|w|^2}{|1-\bar{w}u|^2}$$

is the Poisson kernel.

Necessity: if $f \in \mathcal{Q}_p$, then $\|f\|_{\mathcal{Q}_p,*} < \infty$. Arbitrarily pick a subarc I of $\mathbf{T}$. If $I \neq \mathbf{T}$, then we choose a point $w \in \mathbf{D}\setminus\{0\}$ such that $w/|w|$ and $2\pi(1-|w|)$ are the center and arclength of I, respectively. If $I = \mathbf{T}$, then we take $w = 0$. With such a point w, as well as the inequality $\cos t \geq 1 - 2^{-1}t^2$ for $t \in (-\infty, \infty)$, we get that for $u \in I$,

$$P_w(u) \geq \frac{1}{1-|w|} \approx \frac{1}{|I|}. \tag{6.2}$$

Corollary 1.1.1, Lemma 6.1.1 and an application of (6.2) to (6.1) show

$$\|f\|_{\mathcal{Q}_p,*} \preceq \sup_{u\in\mathbf{D}} \|f\circ\sigma_u\|_{\mathcal{D}_p,*} \preceq \sup_{u\in\mathbf{D}} \|f\circ\sigma_u\|_{\mathcal{D}_p} \preceq \|f\|_{\mathcal{Q}_p} < \infty.$$

Sufficiency: if $\|f\|_{\mathcal{Q}_p,*} < \infty$ then $f \in \mathcal{Q}_p$. To each point $w \in \mathbf{T}\setminus\{0\}$ we associate the subarc I_w with center $w/|w|$ and arclength $2\pi(1-|w|)$. For $w=0$, we set $I_w = \mathbf{T}$. Also, we set

$$I^n = 2^n I_w, \quad n = 0, 1, ..., N-1,$$

where N is the smallest integer such that $2^N|I_w| \geq 2\pi$. And then, we put $I^N = \mathbf{T}$.

Through the help of the elementary inequality $\cos t \leq 1 - 2\pi^{-2}t^2$ for $t \in [-\pi, \pi]$, we know that for every point $u \in \mathbf{T}$,

$$P_w(u) \preceq \frac{1}{1-|w|}. \tag{6.3}$$

Furthermore, for $u \in \mathbf{T}\setminus I^n$ we have

$$P_w(u) \preceq \frac{1}{2^{2n}|w||I_w|}.$$

In the sequel, we may assume $|w| \geq 1/2$, otherwise, the result is obviously true. Therefore, if $u \in I^{n+1} \setminus I^n$, then

$$P_w(u) \preceq \frac{1}{2^{2n}|I_w|}. \tag{6.4}$$

With the above notations, we break $\|f \circ \sigma_w\|^2_{\mathcal{D}_p,*}$ of (6.1) into two parts.

$$\frac{\|f \circ \sigma_w\|^2_{\mathcal{D}_p,*}}{(2\pi)^{\frac{p}{2}}} = \int_{\mathbf{T}} \left(\int_{I_w} + \sum_{n=0}^{N-1} \int_{I^{n+1} \setminus I^n} \right) \frac{|f(u) - f(v)|^2}{|u-v|^{2-p}} \frac{|du||dv|}{\left(P_w(u)P_w(v)\right)^{-p/2}}$$

$$= \int_{\mathbf{T}} \int_{I_w} \left(\cdots \right) + \sum_{n=0}^{N-1} \int_{\mathbf{T}} \int_{I^{n+1} \setminus I^n} \left(\cdots \right) = X_1 + X_2.$$

For convenience, we recall that f_I stands for the average of f on the arc $I \subseteq \mathbf{T}$, namely,

$$f_I = \frac{1}{|I|} \int_I f(z)|dz|.$$

By (6.3), (6.4), the definition of BMO (see Chapter 1), the triangle inequality and the following identity

$$\frac{1}{|I|} \int_I |f(z) - a|^2 |dz| = \frac{1}{|I|} \int_I |f(z) - f_I|^2 |dz| + |f_I - a|^2, \quad a \in \mathbf{C},$$

we have

$$X_1 = \int_{I_w} \int_{I_w} \left(\cdots \right) + \sum_{n=0}^{N-1} \int_{I^{n+1} \setminus I^n} \int_{I_w} \left(\cdots \right)$$

$$\preceq \|f\|^2_{\mathcal{Q}_p,*} + \sum_{n=1}^{N-1} \frac{1}{(2^{2n}|I_w|)^p} \int_{I^{n+1} \setminus I^n} \int_{I_w} \frac{|f(u) - f(v)|^2}{|u-v|^{2-p}} |du||dv|$$

$$\preceq \|f\|^2_{\mathcal{Q}_p,*} + \sum_{n=1}^{N-1} \frac{1}{(2^{2n}|I_w|)^2} \int_{I^{n+1} \setminus I^n} \int_{I_w} |f(u) - f(v)|^2 |du||dv|$$

$$\preceq \|f\|^2_{\mathcal{Q}_p,*} + \left(\sum_{n=1}^{\infty} \frac{1}{2^n} + \sum_{n=1}^{\infty} \frac{n^2}{2^n} \right) \|f\|^2_{BMO}$$

$$\preceq \|f\|^2_{\mathcal{Q}_p,*}.$$

When handling X_2, we omit the integrated functions (for simplicity), and use the same manner as dominating X_1 to obtain

$$\begin{aligned}
X_2 &= \sum_{n=0}^{N-1}\int_{I_w}\int_{I^{n+1}\setminus I^n} + \sum_{n=0}^{N-1}\sum_{m=0}^{N-1}\int_{I^{n+1}\setminus I^n}\int_{I^{m+1}\setminus I^m} \\
&\preceq \|f\|^2_{\mathcal{Q}_p,*} + \sum_{m=0}^{N-1}\int_{I^1\setminus I_w}\int_{I^{m+1}\setminus I^m} + \sum_{n=1}^{N-1}\sum_{m=0}^{N-1}\int_{I^{n+1}\setminus I^n}\int_{I^{m+1}\setminus I^m} \\
&\preceq \|f\|^2_{\mathcal{Q}_p,*} + \sum_{n=1}^{N-1}\Big(\sum_{m<n} + \sum_{m\geq n}\Big)\int_{I^{n+1}\setminus I^n}\int_{I^{m+1}\setminus I^m} \\
&\preceq \|f\|^2_{\mathcal{Q}_p,*} + \left(\sum_{n=1}^{\infty}\frac{n^2}{2^n} + \sum_{n=1}^{\infty}\frac{1}{2^{pn}}\right)\|f\|^2_{BMO} \\
&\preceq \|f\|^2_{\mathcal{Q}_p,*}.
\end{aligned}$$

Combining the estimates of X_1 and X_2, as well as using Corollary 1.1.1 and Lemma 6.1.1, we reach

$$\|f\|_{\mathcal{Q}_p} \preceq \sup_{w\in\mathbf{D}}\|f\circ\sigma_w\|_{\mathcal{D}_p} \preceq \sup_{w\in\mathbf{D}}\|f\circ\sigma_w\|_{\mathcal{D}_p,*} \preceq \|f\|_{\mathcal{Q}_p,*} < \infty,$$

which concludes the proof.

6.2 Weight Condition

By a careful checking with Theorem 1.1.1 and its proof, we find that the weight $(1-|\sigma_w(z)|^2)^p$ can be replaced by its reflection $(|\sigma_w(z)|^{-2}-1)^p$ in case $p \in (0,1)$. More precisely, we have the following theorem.

Theorem 6.2.1. *Let $p \in (0,1)$ and $f \in \mathcal{H}$. Then $f \in \mathcal{Q}_p$ if and only if*

$$\sup_{w\in\mathbf{D}}\int_{\mathbf{D}}|f'(z)|^2(|\sigma_w(z)|^{-2}-1)^p dm(z) < \infty.$$

Proof. It suffices to prove that for $w \in \mathbf{D}$, the left-hand side integral and $F_p(f,w))^2$ are controlled by each other, namely,

$$\int_{\mathbf{D}}|f'(z)|^2(|\sigma_w(z)|^{-2}-1)^p dm(z) \approx (F_p(f,w))^2.$$

By a change of variables, it is enough to check the last equivalence for $w = 0$. In other words,

$$\int_0^1 (M_2(f',r))^2(r^{-2}-1)^p r\,dr \approx \int_0^1 (M_2(f',r))^2\left(1-r^2\right)^p r\,dr$$

which is equivalent to

$$\int_0^1 (M_2(f',r))^2\left(1-r^2\right)^p r^{1-2p}\,dr \approx \int_0^1 (M_2(f',r))^2\left(1-r^2\right)^p r\,dr.$$

This, however, follows without difficulty, since $p \in (0,1)$ and $M_2(f,r)$ is a non-decreasing function of r.

For $p \in (0,1)$, $w \in \mathbf{D}$ and $z \in \mathbf{C}$, let

$$U_w(z) = U_{w,p}(z) = \left|1 - \frac{1}{|\sigma_w(z)|^2}\right|^p = \frac{(1-|w|^2)^p \left||z|^2 - 1\right|^p}{|z-w|^{2p}}.$$

This notation, together with Theorem 6.2.1, leads to a consideration of the weighted norm inequalities.

Given a nonnegative weight function $\omega \in L^1_{loc}(\mathbf{C})$. For $p > 1$ and any Euclidean disk Δ in $\mathbf{C}$, let

$$A_p(\omega, \Delta) = \left(\frac{1}{m(\Delta)}\int_\Delta \omega dm\right)\left(\frac{1}{m(\Delta)}\int_\Delta \left(\frac{1}{\omega}\right)^{1/(p-1)} dm\right)^{p-1}.$$

We say that ω is an A_p-weight provided

$$A_p(\omega) = \sup_{\Delta \subset \mathbf{C}} A_p(\omega, \Delta) < \infty,$$

where the supremum ranges over all Euclidean disks Δ in $\mathbf{C}$ with the area $m(\Delta)$.

Theorem 6.2.2. *Let $p \in (0,1)$. Then there exists a positive constant C such that $U_{w,p}$ is an A_2-weight with $A_2(U_{w,p}) \le C$ for every $w \in \mathbf{D}$.*

Proof. Fix $w \in \mathbf{D}$ and put

$$V_w(z) = V_{w,p}(z) = \frac{\left||z|^2 - 1\right|^p}{|z-w|^{2p}}, \quad z \in \mathbf{C}.$$

It is clear that

$$A_2(U_w) = A_2(U_{w,p}) = A_2(V_{w,p}) = A_2(V_w), \quad w \in \mathbf{D}.$$

So, we can write $V_w(z) = W(z)Y_w(z)$, where

$$W(z) = W_p(z) = \left||z|^2 - 1\right|^p, \quad Y_w(z) = Y_{w,p}(z) = |z-w|^{-2p}.$$

It is well known (see e.g. [119, p.218]) that, since $p \in (0,1)$, the weight $Y_0(z) = |z|^{-2p}$ satisfies the A_q-condition for all $q > 1$. Since the Y_w are translates of Y_0, it follows that for every $q > 1$ there exists a constant $C_q > 0$ (depending on q) such that

$$A_q(Y_w) \le C_q \tag{6.5}$$

holds for all $w \in \mathbf{D}$. Take and fix $r \in (1, 1/p)$, and let Δ be any disk in $\mathbf{C}$. Then, for $w \in \mathbf{D}$ and $r' = r/(r-1)$, we have

$$\begin{aligned}
A_2(V_w, \Delta) &= \left(\frac{1}{m(\Delta)}\int_\Delta V_w(z)\, dm(z)\right)\left(\frac{1}{m(\Delta)}\int_\Delta \frac{dm(z)}{V_w(z)}\right) \\
&= \left(\frac{1}{m(\Delta)}\int_\Delta W(z)Y_w(z)\, dm(z)\right)\left(\frac{1}{m(\Delta)}\int_\Delta \frac{dm(z)}{W(z)Y_w(z)}\right) \\
&\le \left(\sup_{z\in\Delta} W(z)\right)\left(\frac{1}{m(\Delta)}\int_\Delta Y_w(z) dm(z)\right) \\
&\times \left(\frac{1}{m(\Delta)}\int_\Delta \frac{dm(z)}{(W(z))^r}\right)^{1/r}\left(\frac{1}{m(\Delta)}\int_\Delta \frac{dm(z)}{(Y_w(z))^{r'}}\right)^{1/r'}.
\end{aligned}$$

Now it can be easily proved by a direct calculation that

$$\left(\sup_{z\in\Delta} W(z)\right)\left(\frac{1}{m(\Delta)}\int_\Delta \frac{dm(z)}{(W(z))^r}\right)^{1/r} \preceq 1.$$

Consequently, we have

$$A_2(V_w,\Delta) \preceq \left(\frac{1}{m(\Delta)}\int_\Delta Y_w(z)\,dm(z)\right)\left(\frac{1}{m(\Delta)}\int_\Delta \frac{dm(z)}{(Y_w(z))^{r'}}\right)^{1/r'}.$$

Next, we set $q = 1 + 1/r'$, and $q' = q/(q-1)$ so that $q/q' = 1/r'$ and rewrite the last inequality as

$$A_2(V_w,\Delta) \preceq \left(\frac{1}{m(\Delta)}\int_\Delta Y_w(z)dm(z)\right)\left(\frac{1}{m(\Delta)}\int_\Delta \frac{dm(z)}{(Y_w(z))^{q'/q}}\right)^{q/q'}.$$

This, together with (6.5), implies $A_2(V_w,\Delta) \preceq C_q$ for every disk $\Delta \subset \mathbf{C}$ and every $w \in \mathbf{D}$. We are done.

Each A_2-weight induces an L^2-bounded Calderón-Zygmund operator. To be precise, we say that a kernel k of the type $k(z) = \Omega(z)|z|^{-2}$ $(z \in \mathbf{C})$ is Calderón-Zygmund kernel if Ω is homogeneous of degree zero: $\Omega(cz) = \Omega(z)$ for $z \in \mathbf{C}$ and $c > 0$, but also has mean value zero on the unit circle: $\int_{\mathbf{T}} \Omega(z)|dz| = 0$. For such a kernel we consider the singular integral

$$\begin{aligned}(Kf)(z) &= \text{p.v.}\int_{\mathbf{C}} f(z-w)\Omega(w)|w|^{-2}dm(w)\\ &= \lim_{\epsilon\to 0}\int_{|z-w|>\epsilon} f(z-w)\Omega(w)|w|^{-2}dm(w),\end{aligned}$$

and the operator $K : f \to Kf$ is called a Calderón-Zygmund operator.

The well known result on the boundedness (cf. [44] for example) of Calderón-Zygmund operator is to say that if ω is an A_2-weight with $A_2(\omega) \preceq C$ and if Ω satisfies also a "Dini-type" condition, i.e.,

$$\int_0^1 \sup\{|\Omega(z_1) - \Omega(z_2)| : z_1, z_2 \in \mathbf{T},\ \ |z_1 - z_2| < \rho\}\rho^{-1}d\rho < \infty,$$

then one has

$$\int_{\mathbf{C}} |(Kg)(z)|^2\omega(z)\,dm(z) \preceq \int_{\mathbf{C}} |g(z)|^2\omega(z)\,dm(z), \quad g \in L^2(\omega). \tag{6.6}$$

Here it is worth pointing out that the constant before the right-hand side integral of (6.6) depends only on both C (the A_2-bound of ω above) and $\|K\|_{L^2\to L^2}$ (the norm of K in the unweighted L^2-space).

6.3 Pseudo-holomorphic Continuation

In this section, $\bar{\mathbf{D}}$ denotes the closed unit disk and $\mathbf{D}^c$ the region $\mathbf{C} \setminus \bar{\mathbf{D}}$. Put $z^\star = 1/\bar{z}$ for $z \in \mathbf{C} \setminus \{0\}$. For $z = x + iy$, let

$$\frac{\partial}{\partial \bar{z}} = \frac{1}{2}\left(\frac{\partial}{\partial x} + i\frac{\partial}{\partial y}\right); \quad \frac{\partial}{\partial z} = \frac{1}{2}\left(\frac{\partial}{\partial x} - i\frac{\partial}{\partial y}\right).$$

be the Cauchy-Riemann operators.

Theorem 6.3.1. *Let $p \in (0,1)$ and $f \in \cap_{q\in(0,\infty)} H^q$. Then $f \in \mathcal{Q}_p$ if and only if there exists a function $F \in C^1(\mathbf{D}^c)$ satisfying:*
(i) *$F(z) = O(1)$ as $z \to \infty$,*
(ii) *$\lim_{r\to 1^+} F(r\zeta) = f(\zeta)$ a.e. on $\mathbf{T}$ and in $L^q(\mathbf{T})$ for all $q \in [1,\infty)$,*
(iii)

$$\sup_{w\in\mathbf{D}} \int_{\mathbf{D}^c} \left|\frac{\partial F(z)}{\partial \bar{z}}\right|^2 \left(|\phi_w(z)|^2 - 1\right)^p dm(z) < \infty.$$

Proof. Necessity: let $f \in \mathcal{Q}_p$, then we show that the above F exists. Set $F(z) = f(z^\star)$ for $z \in \mathbf{D}^c$. It is clear that F is C^1 on $\mathbf{D}^c$ and satisfies (i) and (ii). Now letting $w \in \mathbf{D}$, making the change of variables $z = u^\star$ in the integral which appears in (iii) and noticing that $|\partial F(z)/\partial \bar{z}| = |f'(z^\star)||z^\star|^2$, we obtain

$$\int_{\mathbf{D}^c} \left|\frac{\partial F(z)}{\partial \bar{z}}\right|^2 \left(|\phi_w(z)|^2 - 1\right)^p dm(z) = \int_{\mathbf{D}} |f'(u)|^2 \left(|\phi_w(u)|^{-2} - 1\right)^p dm(u).$$

Then (iii) follows from Theorem 6.2.1.

Sufficiency: suppose that there is an $F \in C^1(\mathbf{D}^c)$ such that the above conditions (i), (ii) and (iii) hold, we verify $f \in \mathcal{Q}_p$. Fix $z \in \mathbf{D}$ and $R > 1$. In view of (ii), the Cauchy-Green formula applied to the function that equals f in $\mathbf{D}$ and F in $\mathbf{D}^c$ gives

$$f(z) = \frac{1}{2\pi i}\int_{|\zeta|=R} \frac{F(\zeta)}{\zeta - z}\, d\zeta - \frac{1}{\pi}\int_{1<|\xi|<R} \frac{\partial F(\xi)}{\partial \bar{\xi}} \frac{dm(\xi)}{\xi - z}.$$

Differentiating this equation and observing that the arising contour integral is $O(1/R)$ as $R \to \infty$, we obtain

$$f'(z) = -\frac{1}{\pi}\int_{\mathbf{D}^c} \frac{\partial F(\xi)}{\partial \bar{\xi}} \frac{dm(\xi)}{(\xi - z)^2}.$$

Put

$$\Phi(z) = \begin{cases} \partial F(z)/\partial \bar{z}, & \text{if } z \in \mathbf{D}^c, \\ 0, & \text{if } z \in \mathbf{D}. \end{cases}$$

Let now T be the Beurling transformation as follows:

$$(Tg)(z) = \text{p.v.} \int_{\mathbf{C}} \frac{g(\xi)}{(\xi - z)^2}\, dm(\xi).$$

It is well known that such a transformation is a Calderón-Zygmund operator.

Using the previous formulas, we see

$$f'(z) = -\pi^{-1}(T\Phi)(z), \quad z \in \mathbf{D}.$$

By Theorems 6.2.1 and 6.2.2 and (6.6) – the boundedness of Calderón-Zygmund operator, we get that for $w \in \mathbf{D}$,

$$\begin{aligned}
(F_p(f,w))^p &\preceq \int_{\mathbf{D}} |f'(z)|^2(|\sigma_w(z)|^{-2}-1)^p dm(z) \\
&= \int_{\mathbf{D}} |f'(z)|^2 U_{w,p}(z)\, dm(z) \\
&\approx \int_{\mathbf{D}} |(T\Phi)(z)|^2 U_{w,p}(z)\, dm(z) \\
&\preceq \int_{\mathbf{C}} |(T\Phi)(z)|^2 U_{w,p}(z)\, dm(z) \\
&\preceq \int_{\mathbf{C}} |\Phi(z)|^2 U_{w,p}(z)\, dm(z) \\
&\preceq \int_{\mathbf{D}^c} \left|\frac{\partial F(z)}{\partial \bar{z}}\right|^2 U_{w,p}(z)\, dm(z) \\
&\preceq \int_{\mathbf{D}^c} \left|\frac{\partial F(z)}{\partial \bar{z}}\right|^2 (|\sigma_w(z)|^2-1)^p\, dm(z).
\end{aligned}$$

Therefore, the condition (iii) and Theorem 1.1.1 complete the proof.

6.4 $\mathcal{K}$-property

To present some consequences of Theorem 6.3.1, we introduce further terminology. We recall first that, given a function $v \in L^\infty(\mathbf{T})$, the associated Toeplitz operator T_v is defined by

$$(T_v f)(z) = \frac{1}{2\pi i}\int_{\mathbf{T}} \frac{v(\zeta)f(\zeta)}{\zeta - z} d\zeta, \quad f \in H^1, z \in \mathbf{D}.$$

A subspace X of H^1 is said to have the $\mathcal{K}$-property provided $T_{\bar{\psi}}(X) \subseteq X$ for any $\psi \in H^\infty$. Moreover, if $f/B \in X$ whenever $f \in X$ and B is an inner function with $f/B \in H^1$, then we say that X has the $\mathcal{F}$-property. It is known that the $\mathcal{K}$-property implies the $\mathcal{F}$-property: indeed, if $f \in H^1$, B is inner and $f/B \in H^1$ then $f/B = T_{\bar{B}} f$.

Theorem 6.4.1. *Let $p \in (0,1)$. Then $\mathcal{Q}_p$ has the $\mathcal{K}$-property.*

Proof. Suppose $f \in \mathcal{Q}_p$ and $\psi \in H^\infty$. We have to show that $g = T_{\bar{\psi}} f$ is necessarily in $\mathcal{Q}_p$.

Since g is the orthonormal projection of $f\bar{\psi}$ onto H^2, one has $f\bar{\psi} = g + \bar{h}$ for some $h \in H_0^2$, the space of $f \in H^2$ with $f(0) = 0$. Thus,

$$g = f\bar{\psi} - \bar{h} \ \ a.e. \ \ on \ \ \mathbf{T}. \tag{6.7}$$

Now, since $f \in \mathcal{Q}_p$, Theorem 6.3.1 tells us that there is a function $F \in C^1(\mathbf{D}^c)$ obeying (i), (ii) and (iii). Further, we set, for $z \in \mathbf{D}^c$,

$$\Psi(z) = \overline{\psi(z^*)}; \quad H(z) = \overline{h(z^*)}$$

and finally

$$G(z) = F(z)\Psi(z) - H(z).$$

In what follows, we claim that $G|_{\mathbf{T}} = g$ (the boundary values are taken in the sense of radial convergence a.e. on $\mathbf{T}$ and in each L^q with $q < \infty$) and

$$\left|\frac{\partial G(z)}{\partial \bar{z}}\right| \leq \|\psi\|_{H^\infty} \left|\frac{\partial F(z)}{\partial \bar{z}}\right|, \quad z \in \mathbf{D}^c. \tag{6.8}$$

In fact, the equation $G|_{\mathbf{T}} = g$ follows from (6.7) and the facts:

$$F|_{\mathbf{T}} = f, \quad \Psi|_{\mathbf{T}} = \bar{\psi}, \quad H|_{\mathbf{T}} = \bar{h},$$

while (6.8) holds because Ψ and H are holomorphic in $\mathbf{D}^c$, and so

$$\frac{\partial G(z)}{\partial \bar{z}} = \Psi(z)\frac{\partial F(z)}{\partial \bar{z}}, \quad z \in \mathbf{D}^c.$$

Since G is obviously C^1 on $\mathbf{D}^c$ and bounded at ∞, we now conclude from $G|_{\mathbf{T}} = g$ and (6.8) that the analogies of (i), (ii) and (iii) hold true with G and g in place of F and f. Another application of Theorem 6.3.1 yields $g \in \mathcal{Q}_p$, as desired.

Of course, this theorem in turn implies that each $\mathcal{Q}_p$ also enjoys the $\mathcal{F}$-property.

Notes

6.1 Most of Section 6.1 is from the papers by Nicolau-Xiao [95] and Xiao [136]. For an alternate proof of Theorem 6.1.1, see Essén-Xiao [63].

6.2 Holding the properties of the boundary value functions of $\mathcal{Q}_p$, we may consider the pseudo-holomorphic extension of the $\mathcal{Q}_p$-functions across the unit circle $\mathbf{T}$. This idea is closely related to Dyakonov-Girela's work [56] whose main contents are adapted as the above Sections 6.2, 6.3 and 6.4. We refer to Dyn'kin's paper [57] for similar descriptions of the classical smoothness spaces, as well as to Dyakonov's work [53] for other important applications of the pseudo-holomorphic extension method. In addition, Theorem 6.2.1 has a sort of higher dimension version; see Gürlebeck-Kähler-Shapiro-Tovar [73] for details.

6.3 Though all the results in this chapter live on the condition $p \in (0,1)$, it is readily seen from Havin [74] that the endpoint spaces $\mathcal{D}$ and BMOA enjoy the $\mathcal{K}$-property. Here it is also worth pointing out that the Bloch space $\mathcal{B}$ fails at this property; see Anderson [6]. For the setting of the weighted Dirichlet spaces, see also Dyakonov [52], [54] and Rabindranathan [102].

7. Representation via $\bar{\partial}$-equation

The Fefferman-Stein decomposition theorem for BMOA is to say that every BMOA-function f can be written as the sum f_1+if_2 where $f_1, f_2 \in \mathcal{H}$ and $\mathrm{Re}f_j \in L^\infty(\mathbf{T})$. The main aim of this chapter is to extend this result to $\mathcal{Q}_p$, $p \in (0,1)$. This aim will be realized via: introducing $\mathcal{Q}_p(\mathbf{T})$, the non-holomorphic version of $\mathcal{Q}_p$; finding the $\mathcal{Q}_p(\mathbf{T}) \cap L^\infty(\mathbf{T})$-solutions to the $\bar{\partial}$-equation; and presenting the Fefferman-Stein type decomposition for $\mathcal{Q}_p$. As certain applications of the $\bar{\partial}$-equation, we give the corona theorems for both $\mathcal{Q}_p \cap H^\infty$ and $\mathcal{Q}_p$, and then show the interpolation theorem for $\mathcal{Q}_p \cap H^\infty$.

7.1 Harmonic Extension

Theorem 6.1.1 induces the following concept.

Definition 7.1.1. *Let $p \in (0,1)$ and $f \in L^2(\mathbf{T})$. Then we say $f \in \mathcal{Q}_p(\mathbf{T})$ provided*

$$\|f\|^2_{\mathcal{Q}_p,*} = \sup_{I \subseteq \mathbf{T}} |I|^{-p} \int_I \int_I \frac{|f(\zeta)-f(\eta)|^2}{|\zeta-\eta|^{2-p}} |d\zeta||d\eta| < \infty,$$

where the supremum is taken over all arcs $I \subseteq \mathbf{T}$.

For $f \in L^1(\mathbf{T})$, let $\hat{f}$ be the harmonic extension of f to $\mathbf{D}$, i.e.,

$$\hat{f}(z) = \int_{\mathbf{T}} f(\zeta) d\mu_z(\zeta), \quad z \in \mathbf{D},$$

where μ_z is still the Poisson measure defined in Chapter 5. Below is the Stegenga's lemma.

Lemma 7.1.1. *Let I and J be two arcs on $\mathbf{T}$ centered at $\zeta_0 = e^{is_0}$ with $|J| \geq 3|I|$. If $f \in L^2(\mathbf{T})$. Then*

$$\int_{S(I)} |\nabla \hat{f}(z)|^2 (1-|z|^2)^p dm(z) \preceq \int_J \int_J \frac{|f(e^{it})-f(e^{is})|^2}{|e^{it}-e^{is}|^{2-p}} dtds + |I|^{2+p} \left(\int_{|t| \geq 2|J|/3} \frac{|f(e^{i(t+s_0)})-f_J|}{t^2} dt \right)^2.$$

Proof. Assume without loss of generality that ζ_0 is 1 and ϕ is a function with $0 \le \phi \le 1$ such that $\phi = 1$ on $J/3$, supp$\phi \subset 2J/3$, and $|\phi(z) - \phi(w)| \preceq |z - w|/|J|$ for all $z, w \in \mathbf{T}$. Writing $\psi = 1 - \phi$, we then have

$$f = f_J + (f - f_J)\phi + (f - f_J)\psi = f_1 + f_2 + f_3.$$

In the integral with the gradient square, f_1 contributes nothing since it is constant. Accordingly, $|\nabla \hat{f}|^2$ is dominated by $|\nabla \hat{f}_3|^2$ and $|\nabla \hat{f}_2|^2$.

For $z = re^{i\theta}$ in the Carleson box $S(I)$,

$$|\nabla \hat{f}_3(re^{i\theta})| \preceq \int_0^{2\pi} \frac{|f_3(e^{it})|}{(1-r)^2 + (\theta - t)^2} dt \preceq \int_{|t| \ge 2|J|/3} |f(e^{it}) - f_J| \frac{dt}{t^2}$$

and therefore by the elementary estimates,

$$\int_{S(I)} |\nabla \hat{f}_3(z)|^2 (1 - |z|^2)^p dm(z) \preceq |I|^{2+p} \left(\int_{|t| \ge 2|J|/3} |f(e^{it}) - f_J| \frac{dt}{t^2} \right)^2.$$

Now for the integral over $S(I)$ of $|\nabla \hat{f}_2|^2$ we replace $S(I)$ with $\mathbf{D}$, and using the Fourier series of f_2 (see also the proof of Lemma 6.1.1) we obtain

$$\begin{aligned}\int_{S(I)} |\nabla \hat{f}_2(z)|^2 (1 - |z|^2)^p dm(z) &\preceq \int_{\mathbf{T}} \int_{\mathbf{T}} \frac{|f_2(\zeta) - f_2(\eta)|^2}{|\zeta - \eta|^{2-p}} |d\zeta||d\eta| \\ &= \int_{\zeta, \eta \in J} + \int_{\zeta \notin J, \eta \in 3J/4} + \int_{\eta \notin J, \zeta \in 3J/4} \\ &= T_1 + T_2 + T_3.\end{aligned}$$

For T_1, we observe that for $\zeta, \eta \in \mathbf{T}$,

$$|f_2(\zeta) - f_2(\eta)| \preceq |f(\zeta) - f(\eta)| + |J|^{-1} |\zeta - \eta| |f(\eta) - f_J|.$$

Thus, we need only to estimate

$$\begin{aligned}\frac{1}{|J|^2} \int_J \int_J \frac{|f(\eta) - f_J|^2}{|\zeta - \eta|^p} |d\zeta||\eta| &= \frac{1}{|J|^2} \int_J |f(\eta) - f_J|^2 \left(\int_J \frac{|d\zeta|}{|\zeta - \eta|^p} \right) |d\eta| \\ &\preceq \frac{1}{|J|^{1-p}} \int_J |f(\eta) - f_J|^2 |d\eta| \\ &\preceq \int_J \int_J \frac{|f(\zeta) - f(\eta)|^2}{|\zeta - \eta|^{2-p}} |d\zeta||d\eta|,\end{aligned}$$

which gives the estimate for T_1. The T_2 and T_3 terms are handled similarly as the last estimate, using $f_2(\zeta) = 0$ for $\zeta \notin J$. Combining the above inequalities implies the lemma.

Motivated by those characterizations of $\mathcal{Q}_p$ in the previous chapters, we can easily establish the following result.

Theorem 7.1.1. *Let $p \in (0,1)$ and let $f \in BMO(\mathbf{T})$. Then the following conditions are equivalent:*

(i) $f \in \mathcal{Q}_p(\mathbf{T})$.

(ii)

$$\sup_{I\subseteq \mathbf{T}} \int_0^{|I|} \frac{1}{t^{2-p}} \left(\int_I |f(e^{i(\theta+t)}) - f(e^{i\theta})|^2 d\theta \right) dt < \infty.$$

where the supremum is taken over all arcs $I \subseteq \mathbf{T}$.

(iii) $|\nabla \hat{f}(z)|^2 (1-|z|^2)^p dm(z)$ *is a p-Carleson measure.*

(iv)

$$\sup_{w\in\mathbf{D}} \int_{\mathbf{D}} |\nabla \hat{f}(z)|^2 (1-|\sigma_w(z)|^2)^p dm(z) < \infty.$$

(v)

$$\sup_{w\in\mathbf{D}} \int_{\mathbf{D}} |\nabla \hat{f}(z)|^2 (|\sigma_w(z)|^{-2} - 1)^p dm(z) < \infty.$$

(vi)

$$\sup_{w\in\mathbf{D}} \int_{\mathbf{D}} \left(\int_{\mathbf{T}} |f(\zeta) - \hat{f}(z)|^2 d\mu_z(\zeta) \right) (1-|\sigma_w(z)|^2)^p d\lambda(z) < \infty.$$

(vii)

$$\sup_{w\in\mathbf{D}} \int_{\mathbf{D}} \left(\widehat{|f|^2}(z) - |\hat{f}(z)|^2 \right) (1-|\sigma_w(z)|^2)^p d\lambda(z) < \infty.$$

Proof. It suffices to verify that (i), (ii) and (iii) are equivalent.

Let (iii) hold. Assume without loss of generality that I is an interval $(0,|I|)$ with $|I| \le \frac{1}{4}$. If $t \in (0,|I|)$ then, Minkowski's inequality is applied to imply

$$\begin{aligned}\left(\int_{3I} |f(e^{i(v+t)}) - f(e^{iv})|^2 dv \right)^{\frac{1}{2}} &\le 2 \int_{1-t}^{1} \left(\int_{4I} \left| \frac{\partial \hat{f}}{\partial n}(ue^{is}) \right|^2 ds \right)^{\frac{1}{2}} du \\ &\quad + 2 \int_0^t \left(\int_{4I} \left| \frac{\partial \hat{f}}{\partial \theta}((1-t)e^{is}) \right|^2 ds \right)^{\frac{1}{2}} du \\ &= Int_1 + Int_2,\end{aligned}$$

where $\partial f/\partial n$ and $\partial f/\partial \theta$ are the directional derivative of f relative to the radius and the argument, respectively.

Making use of Hardy's inequality (cf. [118, p.272]), we obtain

$$\int_0^{|I|}\frac{(Int_1)^2}{t^{2-p}}dt \preceq \int_0^{|I|} t^p\left(\int_{4I}\left|\frac{\partial\hat{f}}{\partial n}((1-t)e^{is})\right|^2 ds\right)dt$$
$$\preceq \int_{S(4I)} |\nabla\hat{f}(z)|^2(1-|z|)^p dm(z).$$

Meanwhile, Int_2 obeys

$$\int_0^{|I|}\frac{(Int_1)^2}{t^{2-p}}dt \preceq \int_0^{|I|} t^p\left(\int_{4I}\left|\frac{\partial\hat{f}}{\partial \theta}(re^{is})\right|^2 ds\right)dt$$
$$\preceq \int_{S(4I)} |\nabla\hat{f}(z)|^2(1-|z|)^p dm(z).$$

Putting these inequalities in order, we see that (iii) implies (ii).

Suppose that (ii) is valid, to prove (i), it suffices to consider small subarc I of $\mathbf{T}$ (say $|I| \le 1/4$). Of course, I may be assumed to equal a subinterval (a, b) of $[0, 2\pi)$, and hence some elementary calculations give

$$\int_I\int_I \frac{|f(e^{is})-f(e^{it})|^2}{|e^{is}-e^{it}|^{2-p}}dsdt \preceq \int_a^b\int_{a<s+t<b}\frac{|f(e^{i(s+t)})-f(e^{is})|^2}{|t|^{2-p}}dtds$$
$$= \int_a^b\left(\int_{a-s}^0+\int_0^{b-s}\right)\frac{|f(e^{i(s+t)})-f(e^{is})|^2}{|t|^{2-p}}dtds$$
$$\preceq \int_0^{b-a}\frac{1}{t^{2-p}}\left(\int_I |f(e^{i(s+t)})-f(e^{is})|^2 ds\right)dt,$$

which implies (i).

Finally, let us prove the implication $(i) \Rightarrow (iii)$. Note that $|e^{is}-e^{it}| \le |I|$ holds for $e^{is}, e^{it} \in I$. So, if (i) holds, then

$$\int_I\int_I |f(e^{is})-f(e^{it})|^2 dsdt \le \|f\|^2_{\mathcal{Q}_{p,*}}|I|^2.$$

Consequently,

$$\int_I |f(e^{is})-f_I|ds \preceq \|f\|_{\mathcal{Q}_{p,*}}|I|.$$

For simplicity, let $J = 3I$, $|I| < 1/3$, and $\zeta_0 = 0$ in Lemma 7.1.1. Since f lies in $BMO(\mathbf{T})$, this implies that for any subarcs I and J of $\mathbf{T}$ satisfying $I \subset J$ and $|J| \le 3|I|$, one has

$$|f_I - f_J|^2 \le \frac{3}{|J|}\int_J |f(e^{is})-f_J|^2 ds \le 3\|f\|^2_{BMO}.$$

Thus

$$\begin{aligned}\int_{|t|\geq|J|/3}|f(e^{it})-f_J|\frac{dt}{t^2} &\leq \sum_{k=1}^{\infty}\int_{3^{k-1}|J|\leq|t|\leq 3^k|J|}|f(e^{it})-f_J|\frac{dt}{t^2}\\ &\preceq \sum_{k=1}^{\infty}\frac{1}{(3^k|J|)^2}\int_{|t|\leq 3^k|J|}|f(e^{it})-f_{3^{k+1}I}|dt\\ &+\sum_{k=1}^{\infty}\frac{|f_{3^{k+1}I}-f_J|}{3^k|J|}\\ &\preceq \left(\sum_{k=1}^{\infty}\frac{1}{3^k|J|}+\sum_{k=1}^{\infty}\frac{k}{3^k|J|}\right)\|f\|_{BMO}\\ &\preceq \frac{\|f\|_{\mathcal{Q}_p,*}}{|I|}.\end{aligned}$$

This, together with Lemma 7.1.1, produces (iii).

From the argument of the above implication $(iii) \Longrightarrow (i)$ it can be readily seen that every $\mathcal{Q}_p(\mathbf{T})$-function admits different extension than the harmonic extension. More precisely, we have

Corollary 7.1.1. *Let $p \in (0,1)$ and F be a function defined on $\bar{\mathbf{D}}$ such that $F \in C^1(\mathbf{D})$ and $F|_{\mathbf{T}} = f$. If $|\nabla F(z)|^2(1-|z|^2)^p dm(z)$ is a p-Carleson measure, then $f \in \mathcal{Q}_p(\mathbf{T})$.*

For $f \in L^1(\mathbf{T})$ let $\tilde{f}$ denote its harmonic conjugate, namely,

$$\tilde{f}(z) = \frac{1}{2\pi}\int_{\mathbf{T}} Im\left(\frac{\zeta+z}{\zeta-z}\right) f(\zeta)|d\zeta|, \quad z \in \mathbf{D}.$$

And for $\zeta \in \mathbf{T}$ put $\tilde{f}(\zeta) = \lim_{r\to 1}\tilde{f}(r\zeta)$, the radial limit function of $\tilde{f}$.

Corollary 7.1.2. *Let $p \in (0,1)$. If $f \in \mathcal{Q}_p(\mathbf{T})$ then $\tilde{f} \in \mathcal{Q}_p(\mathbf{T})$.*

Proof. Since every $f \in BMO(\mathbf{T})$ enjoys the following identity:

$$\int_{\mathbf{T}}|f(\zeta)-\hat{f}(z)|^2 d\mu_z(\zeta) = \int_{\mathbf{T}}|\tilde{f}(\zeta)-\tilde{f}(z)|^2 d\mu_z(\zeta), \quad z \in \mathbf{D},$$

the corollary follows directly from Theorem 7.1.1.

Once P stands for the Szegö projection from $L^2(\mathbf{T})$ onto H^2:

$$Pf(z) = \frac{1}{2\pi}\int_{\mathbf{T}}\frac{f(\zeta)}{1-\bar{\zeta}z}|d\zeta|, \quad f \in L^2(\mathbf{T}), \quad \mathbf{z} \in \mathbf{D},$$

Corollary 7.1.2 has an interesting consequence.

Corollary 7.1.3. *Let $p \in (0,1)$. Then $P : \mathcal{Q}_p(\mathbf{T}) \to \mathcal{Q}_p$ is bounded and surjective.*

Proof. A simple calculation shows

$$\hat f(z) + i\tilde{\hat f}(z) = \frac{1}{2\pi}\int_{\mathbf{T}} \frac{\zeta+z}{\zeta-z} f(\zeta)|d\zeta| = 2(Pf)(z) - \hat f(0), \quad z \in \mathbf{D},$$

which gives the boundedness of P by Corollary 7.1.2. Since $Pf = f$ whenever $f \in \mathcal{Q}_p$, P is onto and hence the proof is complete.

7.2 $\bar\partial$-estimates

Invoking a basic theorem due to Carleson [39], Hörmander [78] proved that if gdm is a 1-Carleson measure, then there exists a function f defined on $\bar{\mathbf{D}}$ such that $\partial f/\partial\bar z = g$ on $\mathbf{D}$, and such that the boundary value function f is in $L^\infty(\mathbf{T})$. Here and hereafter we use the same letter for a function on $\bar{\mathbf{D}}$ and its boundary value function on $\mathbf{T}$. Later, Jones [83] gave two constructive methods to solve the $\bar\partial$-equation. Observe that $L^\infty(\mathbf{T})$ is not a subclass of $\mathcal{Q}_p(\mathbf{T})$, $p \in (0,1)$ (see the examples at the beginning of Chapter 5), so it is of interest to solve the $\bar\partial$-equation with boundary value function in $\mathcal{Q}_p(\mathbf{T}) \cap L^\infty(\mathbf{T})$.

Lemma 7.2.1. *Let μ be a 1-Carleson measure. If for $z \in \bar{\mathbf{D}}$ and $\zeta \in \mathbf{T}$,*

$$K\left(\frac{\mu}{\|\mu\|_{C_1}}, z, \zeta\right) = \frac{1-|\zeta|^2}{\pi(1-\bar\zeta z)(z-\zeta)} \times \exp\left(\int_{|w|\ge|\zeta|}\left(\frac{1+\bar w\zeta}{1-\bar w\zeta} - \frac{1+\bar w z}{1-\bar w z}\right)d|\mu|(w)\right),$$

then

$$S_0(\mu)(z) = \int_{\mathbf{D}} K\left(\frac{\mu}{\|\mu\|_{C_1}}, z, \zeta\right) d\mu(\zeta) \tag{7.1}$$

satisfies $S_0(\mu) \in L^1(\mathbf{D})$ and $\partial S_0(\mu)/\partial\bar z = \mu$ on $\mathbf{D}$ in the sense of distribution. Moreover, if $z \in \mathbf{T}$, then the integral in (7.1) converges absolutely and

$$\sup_{z\in\mathbf{T}}\int_{\mathbf{D}}\left|K\left(\frac{\mu}{\|\mu\|_{C_1}}, z, \zeta\right)\right| d|\mu|(\zeta) < \infty.$$

In particular, $S_0(\mu) \in L^\infty(\mathbf{T})$.

Proof. This is one of Jones' $\bar\partial$-solutions. For completeness, we give a proof. On the one hand, if h is C^∞ and has compact support contained in $\mathbf{D}$, then

$$\int_{\mathbf{C}} \frac{\partial(S_0(\mu)(z)h(z))}{\partial\bar z} dm(z) = -\frac{1}{2i}\int_{\mathbf{T}} S_0(\mu)(z)h(z)dz = 0,$$

and hence

$$\int_{\mathbf{C}} S_0(\mu)(z)\frac{\partial h(z)}{\partial \bar{z}}dm(z) = -\int_{\mathbf{C}} h(z)\frac{\partial S_0(\mu)(z)}{\partial \bar{z}}dm(z).$$

However, by (7.1) and Fubini's theorem, it is easy to see that

$$\begin{aligned}\int_{\mathbf{C}} S_0(\mu)(z)\frac{\partial h(z)}{\partial \bar{z}}dm(z) &= \int_{\mathbf{D}}\left(\int_{\mathbf{C}}\frac{\partial h(z)}{\partial \bar{z}}K\left(\frac{\mu}{\|\mu\|_{C_1}}, z, \zeta\right)dm(z)\right)d\mu(\zeta)\\ &= -\int_{\mathbf{C}} h(z)d\mu(z),\end{aligned}$$

so that $\partial S_0(\mu)(z)/\partial\bar{z} = \mu$ follows by letting h run through the translates of an approximate identity (see also [109, p. 31]).

In fact, the most important is to prove the last claim of the theorem; the other two claims follow easily from the proof given below. By the form of $S_0(\mu)$ it is enough to prove the last claim for the case $\mu \geq 0$ and $\|\mu\|_{C_1} = 1$. We first note that if $w, \zeta \in \mathbf{D}$ and $|w| \geq |\zeta|$, then

$$Re\left(\frac{1+\bar{w}\zeta}{1-\bar{w}\zeta}\right) \leq \frac{2(1-|\zeta|^2)}{|1-\bar{w}\zeta|^2}.$$

We also observe that the normalized reproducing kernel

$$k_\zeta(w) = \frac{(1-|\zeta|^2)^{1/2}}{1-\bar{\zeta}w}$$

obeys $\|k_\zeta\|_{H^2} \leq 2$. Consequently,

$$Re\left(\int_{|w|\geq|\zeta|}\frac{1+\bar{w}\zeta}{1-\bar{w}\zeta}d\mu(w)\right) \leq 2\int_{\mathbf{D}}|k_\zeta(w)|^2 d\mu(w) \preceq \|k_\zeta\|_{H^2}^2 \preceq 1.$$

Fix a point $w \in \mathbf{T}$. Since

$$Re\left(-\frac{1+\bar{w}\zeta}{1-\bar{w}\zeta}\right) = -\frac{2(1-|\zeta|^2)}{|1-\bar{w}\zeta|^2},$$

the proof of Lemma 7.2.1 will follow immediately from

$$II_\mu = \int_{\mathbf{D}}\frac{1-|z\bar{\zeta}|^2}{|1-z\bar{\zeta}|^2}\exp\left(-\int_{|w|\geq|\zeta|}\frac{1-|z\bar{w}|^2}{|1-z\bar{w}|^2}d\mu(w)\right)d\mu(\zeta) \leq 1. \tag{7.2}$$

However, this follows from the integral formula $\int_0^\infty e^{-t}dt = 1$. Suppose for example that $\mu = \sum_{j=1}^N a_j\delta_{\zeta_j}$ is a finite weighted sum of Dirac measures. Let $|\zeta_1| \leq |\zeta_2| \leq \cdots \leq |\zeta_N|$ and put

$$b_j = \frac{a_j(1-|\zeta_j|^2)}{|1-\bar{\zeta}_j w|^2}, \quad w \in \mathbf{T}.$$

Then since $|\sigma_\zeta(w)| = 1$ for $\zeta \in \mathbf{D}$ and $w \in \mathbf{T}$,

$$II_\mu \le \sum_{j=1}^{N} b_j \exp\left(-\sum_{k=1}^{j} b_j\right) < 1,$$

because the last sum is a lower Riemann sum for $\int_0^\infty e^{-t}dt$. Standard measure theoretic arguments now complete the proof of (7.2).

Before reaching the main result of this section, we need another lemma which says that some p-Carleson measures are stable under a special integral operator.

Lemma 7.2.2. *Let $p \in (0,1)$ and define*

$$(Tf)(z) = \int_{\mathbf{D}} \frac{f(w)}{|1 - z\bar{w}|^2} dm(w).$$

If $d\mu(z) = |f(z)|^2(1-|z|^2)^p dm(z)$ is a p-Carleson measure, then $d\nu(z) = |(Tf)(z)|^2(1-|z|^2)^p dm(z)$ is also a p-Carleson measure.

Proof. For the Carleson box $S(I)$, we have

$$\begin{aligned}
\nu(S(I)) &= \int_{S(I)} |(Tf)(z)|^2(1-|z|^2)^p dm(z) \\
&\le \int_{S(I)} (1-|z|^2)^p \left(\left(\int_{S(2I)} + \int_{\mathbf{D}\setminus S(2I)}\right) \frac{|f(w)|}{|1-\bar{w}z|^2} dm(w)\right)^2 dm(z) \\
&\preceq \int_{S(I)} (1-|z|^2)^p \left(\int_{S(2I)} \frac{|f(w)|}{|1-\bar{w}z|^2} dm(w)\right)^2 dm(z) \\
&+ \int_{S(I)} (1-|z|^2)^p \left(\int_{\mathbf{D}\setminus S(2I)} \frac{|f(w)|}{|1-\bar{w}z|^2} dm(w)\right)^2 dm(z) \\
&= Int_3 + Int_4.
\end{aligned}$$

For Int_3, we use Schur's lemma [144, p.42]. Indeed, we consider

$$k(z,w) = \frac{(1-|z|^2)^{p/2}(1-|w|^2)^{-p/2}}{|1-\bar{w}z|^2}$$

and its induced integral operator

$$(Lf)(z) = \int_{\mathbf{D}} f(w)k(z,w)\, dm(w).$$

Taking $\alpha \in (-1, -p/2)$ and applying Lemma 1.4.1, we get

$$\int_{\mathbf{D}} k(z,w)(1-|w|^2)^\alpha\, dm(w) \preceq (1-|z|^2)^\alpha$$

and

$$\int_{\mathbf{D}} k(z,w)(1-|z|^2)^{\alpha}\,dm(z) \preceq (1-|w|^2)^{\alpha}.$$

Therefore the operator L is bounded from $L^2(\mathbf{D})$ to $L^2(\mathbf{D})$. Once the function f in Lf is replaced by $g(w) = (1-|w|^2)^{p/2}|f(w)|1_{S(2I)}(w)$, we have

$$\begin{aligned} Int_3 &\preceq \int_{\mathbf{D}}\left(\int_{\mathbf{D}} g(w)k(z,w)\,dm(w)\right)^2 dm(z) \\ &\preceq \int_{\mathbf{D}} |g(z)|^2\,dm(z) \\ &= \int_{S(2I)} |f(z)|^2(1-|z|^2)^p\,dm(z) \\ &\preceq \|\mu\|_{C_p}|I|^p. \end{aligned}$$

Since $d\mu(z) = |f(z)|^2(1-|z|^2)^p dm(z)$ is a p-Carleson measure, $|f(z)|dm(z)$ is a 1-Carleson measure. In fact, the Cauchy-Schwarz inequality gives that for the Carleson box $S(I)$,

$$\left(\int_{S(I)} |f(z)|dm(z)\right)^2 \preceq |I|^{2-p}\int_{S(I)} |f(z)|^2(1-|z|^2)^p dm(z) \preceq |I|^2\|\mu\|_{C_p}.$$

This deduces

$$\begin{aligned} Int_4 &\preceq \int_{S(I)}(1-|z|^2)^p\left(\sum_{n=1}^{\infty}\int_{S(2^{n+1}I)\setminus S(2^nI)} \frac{|f(w)|}{|1-\bar{w}z|^2}\,dm(w)\right)^2 dm(z) \\ &\preceq \|\mu\|_{C_p}\int_{S(I)}(1-|z|^2)^p\left(\sum_{n=1}^{\infty}\frac{2^{n+1}|I|}{(2^n|I|)^2}\right)^2 dm(z) \\ &\preceq \|\mu\|_{C_p}|I|^p. \end{aligned}$$

These estimates on Int_3 and Int_4 imply that ν is a p-Carleson measure.

Theorem 7.2.1. *Let $p \in (0,1)$. If $|g(z)|^2(1-|z|^2)^p dm(z)$ is a p-Carleson measure, then there is a function f defined on $\bar{\mathbf{D}}$ such that*

$$\frac{\partial f(z)}{\partial \bar{z}} = g(z), \quad z \in \mathbf{D},$$

and such that the boundary value function f belongs to $\mathcal{Q}_p(\mathbf{T}) \cap L^{\infty}(\mathbf{T})$.

Proof. By the hypothesis of Theorem 7.2.1 and the Cauchy-Schwarz inequality, gdm is a 1-Carleson measure. Thus, by Lemma 7.2.1, the function $f = S_0(\mu)$ (where $d\mu = gdm$) is defined on $\bar{\mathbf{D}}$. More importantly, the function $f = S_0(\mu)$ satisfies the equation $\partial f/\partial\bar{z} = g$ on $\mathbf{D}$. Furthermore, the boundary value function

f is in $L^\infty(\mathbf{T})$. However, our aim is to verify that the boundary value function f lies in $\mathcal{Q}_p(\mathbf{T})$, so we must show $\|f\|_{\mathcal{Q}_p,*} < \infty$. For this purpose, let

$$F(z) = \frac{i}{\pi}\int_{\mathbf{D}} \frac{1-|\zeta|^2}{|1-\bar{\zeta}z|^2}$$
$$\times \exp\left(\int_{|w|\geq|\zeta|}\left(\frac{1+\bar{w}\zeta}{1-\bar{w}\zeta} - \frac{1+\bar{w}z}{1-\bar{w}z}\right)|g(w)|dm(w)\right) g(\zeta)dm(\zeta).$$

Observe that $F(z)$ has the same boundary values as $zf(z)$ on $\mathbf{T}$. So, from Corollary 7.1.1 we find it to be sufficient to check that $|\nabla F(z)|^2(1-|z|^2)^p dm(z)$ is a p-Carleson measure. Since gdm is a 1-Carleson measure, by the proof of Lemma 7.2.1 one has

$$Re\left(\int_{|w|\geq|\zeta|}\frac{1+\bar{w}\zeta}{1-\bar{w}\zeta}|g(w)|dm(w)\right) \preceq 1$$

and

$$\int_{\mathbf{D}}\frac{1-|z\bar{\zeta}|^2}{|1-z\bar{\zeta}|^2}\exp\left(-\int_{|w|\geq|\zeta|}\frac{1-|z\bar{w}|^2}{|1-z\bar{w}|^2}|g(w)|dm(w)\right)|g(\zeta)|dm(\zeta) \leq 1,$$

thus

$$|\nabla F(z)| \preceq \int_{\mathbf{D}}\frac{|g(w)|}{|1-\bar{w}z|^2}dm(w). \tag{7.3}$$

Since $|g(z)|^2(1-|z|^2)^p dm(z)$ is a p-Carleson measure, an application of Lemma 7.2.2 to (7.3) produces that $|\nabla F(z)|^2(1-|z|^2)^p dm(z)$ is a p-Carleson measure. The proof is complete.

7.3 Fefferman-Stein Type Decomposition

As is well known, there is a close relation between the $\bar{\partial}$-equation and the Fefferman-Stein decomposition asserting that any $f \in BMO(\mathbf{T})$ can be decomposed into $f = u + \tilde{v}$, where $u, v \in L^\infty(\mathbf{T})$. So, it is not surprising that solving the $\bar{\partial}$-equation with appropriate estimates leads to the following assertion.

Theorem 7.3.1. *Let $p \in (0,1)$ and $f \in L^2(\mathbf{T})$. Then $f \in \mathcal{Q}_p(\mathbf{T})$ if and only if $f = u + \tilde{v}$, where $u, v \in \mathcal{Q}_p(\mathbf{T}) \cap L^\infty(\mathbf{T})$.*

Proof. If $f = u + \tilde{v}$, $u, v \in \mathcal{Q}_p(\mathbf{T}) \cap L^\infty(\mathbf{T})$, then it follows from Corollary 7.1.2 that $\tilde{v} \in \mathcal{Q}_p(\mathbf{T})$ and hence $f \in \mathcal{Q}_p(\mathbf{T})$.

On the other hand, it is enough to consider the case that $f \in \mathcal{Q}_p(\mathbf{T})$ is real-valued. We find immediately that $F = f + i\tilde{f} \in \mathcal{Q}_p(\mathbf{T})$ and its harmonic extension $\hat{F} \in \mathcal{Q}_p$. It turns out from Theorem 7.1.1 that $|\nabla\hat{F}(z)|^2(1-|z|^2)^p\, dm(z)$, and then $|\partial f(z)/\partial\bar{z}|^2(1-|z|^2)^p\, dm(z)$ is a p-Carleson measure. Let $d\mu(z) =$

$(\partial f(z)/\partial\bar{z})dm(z)$ and let $f_\mu(z)$ be the function given by Theorem 7.2.1; then $\partial f_\mu(z)/\partial\bar{z} = \mu$ and $f_\mu \in \mathcal{Q}_p(\mathbf{T}) \cap L^\infty(\mathbf{T})$. Hence $g = f - f_\mu$ is holomorphic and $g \in \mathcal{Q}_p$. Put $u = Ref_\mu$, then $f - u = -\widetilde{Img}$. So $f = u + \widetilde{v}$, where $u = Ref_\mu$ and $v = -Img$ belong to $\mathcal{Q}_p(\mathbf{T}) \cap L^\infty(\mathbf{T})$.

Theorem 7.3.2. *Let $p \in (0,1)$ and $f \in H^2$. Then $f \in \mathcal{Q}_p$ if and only if $f = f_1 + f_2$ where $f_1, f_2 \in \mathcal{H}$ and $Ref_1, Ref_2 \in \mathcal{Q}_p(\mathbf{T}) \cap L^\infty(\mathbf{T})$.*

Proof. This follows immediately from Theorem 7.3.1.

The following result improves Corollary 7.1.3.

Theorem 7.3.3. *Let $p \in (0,1)$. Then the Szegö projection P maps $\mathcal{Q}_p(\mathbf{T}) \cap L^\infty(\mathbf{T})$ onto $\mathcal{Q}_p$.*

Proof. It suffices to show that P is onto. By Theorem 7.3.1, we see that if $f \in \mathcal{Q}_p$ then there are $g, h \in \mathcal{Q}_p(\mathbf{T}) \cap L^\infty(\mathbf{T})$ such that $f = g + \tilde{h}$. This gives

$$f = Pf = Pg + P\tilde{h} = Pg + 2^{-1}(i\tilde{\tilde{h}} + \tilde{h} - \hat{\tilde{h}}(0)) = P(g - ih) + \hat{h}(0) - \hat{\tilde{h}}(0),$$

concluding the proof.

7.4 Corona Data and Solutions

Carleson's corona theorem [38] asserts that if $g_1, g_2, \cdots, g_n$ (corona data) in H^∞ satisfy $\inf_{z\in\mathbf{D}} \sum_{j=1}^n |g_j(z)| \ge \gamma > 0$, then there exist $f_1, f_2, \cdots, f_n$ (corona solutions) in H^∞ such that $\sum_{j=1} f_j g_j = 1$. More is true.

Theorem 7.4.1. *Let $p \in (0,1)$ and $n \in \mathbf{N}$. If $g_1, \cdots, g_n \in \mathcal{Q}_p \cap H^\infty$ satisfy*

$$\gamma = \inf_{z\in\mathbf{D}} \sum_{k=1}^n |g_k(z)| > 0, \tag{7.4}$$

then there exist $f_1, f_2, \cdots, f_n \in \mathcal{Q}_p \cap H^\infty$ such that $\sum_{j=1}^n f_j g_j = 1$.

Proof. The argument will use Theorem 7.2.1 and Wolff's $\bar{\partial}$-approach (of proving Carleson's corona theorem). Suppose $(g_1, \cdots, g_n) \in \mathcal{Q}_p \cap H^\infty \times \cdots \times \mathcal{Q}_p \cap H^\infty$ obeys (7.4). To find $(g_1, \cdots, g_n) \in \mathcal{Q}_p \cap H^\infty \times \cdots \times \mathcal{Q}_p \cap H^\infty$ such that $\sum_{j=1}^n f_j g_j = 1$, let

$$h_k = \frac{\bar{g}_k}{\sum_{j=1}^n |g_j|^2}, \quad k = 1, \cdots, n. \tag{7.5}$$

Then $(h_1, \cdots, h_n)$ is a solution to the equation $\sum_{k=1}^n g_k h_k = 1$. But this solution is not holomorphic. So, like in the H^∞ setting (see [66, pp. 324-325]), we must modify (7.5). Without loss of generality, by the normal family principle we may assume that each g_k is holomorphic on some neighborhood of $\bar{\mathbf{D}}$. Suppose that we can find functions $b_{j,k}$, $1 \le j, k \le n$, defined on $\bar{\mathbf{D}}$ such that

$$\frac{\partial b_{j,k}(z)}{\partial \bar{z}} = h_j(z)\frac{\partial h_k(z)}{\partial \bar{z}}, \quad z \in \mathbf{D},$$

and such that the boundary value functions $b_{j,k}$ are in $\mathcal{Q}_p(\mathbf{T}) \cap L^\infty(\mathbf{T})$. Then

$$f_k = h_k + \sum_{j=1}^{n}(b_{k,j} - b_{j,k})g_j$$

belongs to $\mathcal{Q}_p \cap H^\infty$ and satisfies $\sum_{k=1}^{n} f_k g_k = 1$. Thus we have only to show that these $\bar{\partial}$-equations admit $\mathcal{Q}_p(\mathbf{T}) \cap L^\infty(\mathbf{T})$ solutions. It is enough to deal with an equation $\partial b/\partial \bar{z} = h$, where $b = b_{j,k}$ and $h = h_j \partial h_k/\partial \bar{z}$. Because each g_k is in $\mathcal{Q}_p \cap H^\infty$, $|g_k'(z)|^2(1-|z|^2)^p dm(z)$ is a p-Carleson measure. Also because of

$$|h(z)|^2 \le \frac{2}{\gamma^6}\sum_{k=1}^{n}|g_k'(z)|^2,$$

$|h(z)|^2(1-|z|^2)^p dm(z)$ is a p-Carleson measure. Therefore, with the help of Theorem 7.2.1, we get a function b defined on $\bar{\mathbf{D}}$ such that b satisfies $\partial b/\partial \bar{z} = h$ on $\mathbf{D}$, and such that the boundary value function b lies in $\mathcal{Q}_p(\mathbf{T}) \cap L^\infty(\mathbf{T})$, as desired.

Theorem 7.4.1 can be extended to $\mathcal{Q}_p$ via its multiplier space. To see this, denote by $M(\mathcal{Q}_p)$ the set of pointwise multipliers of $\mathcal{Q}_p$, i.e.,

$$M(\mathcal{Q}_p) = \{f \in \mathcal{Q}_p : M_f g = fg \in \mathcal{Q}_p \textit{ whenever } g \in \mathcal{Q}_p\}.$$

The following conclusion gives a description of $M(\mathcal{Q}_p)$.

Theorem 7.4.2. *Let $p \in (0,\infty)$. If $f \in M(\mathcal{Q}_p)$ then $f \in H^\infty$ and*

$$\|f\|^2_{M(\mathcal{Q}_p)} = \sup_{I \subseteq \mathbf{T}} \frac{\log^2 \frac{2}{|I|}}{|I|^p} \int_{S(I)} |f'(z)|^2(1-|z|)^p dm(z) < \infty, \tag{7.6}$$

where the supremum ranges over all subarcs I of $\mathbf{T}$. Conversely, if $f \in H^\infty$ and $|f'(z)|^2(1-|z|)^p \log^2(1-|z|)dm(z)$ is a p-Carleson measure, then $f \in M(\mathcal{Q}_p)$.

Proof. Let $f \in M(\mathcal{Q}_p)$. Observe that for a fixed $w \in \mathbf{D}$, the function $g_w(z) = \log(2/(1-\bar{w}z))$ belongs to $\mathcal{Q}_p$ with $\sup_{w\in\mathbf{D}} \|g_w\|_{\mathcal{Q}_p} \preceq 1$ (cf. Corollary 3.1.1 (iii)). Then $fg_w \in \mathcal{Q}_p$ with $\|fg_w\| \preceq 1$ for all $w \in \mathbf{D}$. Since any function $g \in \mathcal{Q}_p$ has the following growth (cf. (1.7)):

$$|g(z)| \preceq \|g\|_{\mathcal{Q}_p} \log\frac{2}{1-|z|}, \quad z \in \mathbf{D}, \tag{7.7}$$

this, together with $\|fg_w\| \preceq 1$, gives that

$$|f(z)g_w(z)| \preceq \|fg_w\|_{\mathcal{Q}_p} \log\frac{2}{1-|z|}, \quad z \in \mathbf{D},$$

so that $f \in H^\infty$.

Concerning (7.6), we argue as follows. Because of $f \in M(\mathcal{Q}_p)$, it follows from Theorem 4.1.1 that for the Carleson box $S(I)$,

$$\int_{S(I)} |(fg_w)'(z)|^2(1-|z|)^p dm(z) \preceq \|fg_w\|_{\mathcal{Q}_p}^2 |I|^p,$$

and so that

$$\int_{S(I)} |f'(z)|^2|g_w(z)|^2(1-|z|)^p dm(z) \preceq (\|fg_w\|_{\mathcal{Q}_p}^2 + \|f\|_{H^\infty}^2)|I|^p.$$

Note that if $w = (1-|I|)e^{i\theta}$ and $e^{i\theta}$ is taken as the center of I then for all $z \in S(I)$, $\log 2/|I| \approx |g_w(z)|$. Whence (7.6) is forced to come out.

On the other hand, assume that $f \in H^\infty$ and $|f'(z)|^2(1-|z|)^p \log^2(1-|z|)dm(z)$ is a p-Carleson measure. With the help of (7.7) we deduce that if $g \in \mathcal{Q}_p$ then for the Carleson box $S(I)$,

$$\begin{aligned}\int_{S(I)} (\cdots) &= \int_{S(I)} |(fg)'(z)|^2(1-|z|)^p dm(z)\\ &\preceq \|g\|_{\mathcal{Q}_p}^2 \int_{S(I)} |f'(z)|^2(1-|z|)^p \log^2(1-|z|)dm(z)\\ &\quad + \|f\|_{H^\infty}^2 \int_{S(I)} |g'(z)|^2(1-|z|)^p dm(z),\end{aligned}$$

and hence $fg \in \mathcal{Q}_p$. In other words, $f \in M(\mathcal{Q}_p)$. The proof is complete.

The $\mathcal{Q}_p$, $p \in (0,1)$, corona theorem is formulated below.

Theorem 7.4.3. *Let $p \in (0,1)$ and $(g_1, \cdots, g_n) \in \mathcal{H} \times \mathcal{H} \cdots \times \mathcal{H}$. Also for $(f_1, \cdots, f_n) \in \mathcal{H} \times \mathcal{H} \cdots \times \mathcal{H}$ let*

$$M_{(g_1,\cdots,g_n)}(f_1, \cdots, f_n) = \sum_{k=1}^n f_k g_k.$$

Then $M_{(g_1,g_2,\cdots,g_n)} : \mathcal{Q}_p \times \mathcal{Q}_p \times \cdots \times \mathcal{Q}_p \to \mathcal{Q}_p$ is surjective if and only if $(g_1, g_2, \cdots, g_n) \in M(\mathcal{Q}_p) \times M(\mathcal{Q}_p) \times \cdots \times M(\mathcal{Q}_p)$ satisfies (7.4).

Proof. Suppose that $M_{(g_1,\cdots,g_n)} : \mathcal{Q}_p \times \mathcal{Q}_p \times \cdots \times \mathcal{Q}_p \to \mathcal{Q}_p$ is surjective. Evidently, it is enough to check (7.4). For this, we use the open map theorem to get that to $f \in \mathcal{Q}_p$ there correspond $f_1, f_2, \cdots, f_n \in \mathcal{Q}_p$ with $\|f_k\|_{\mathcal{Q}_p} \preceq \|f\|_{\mathcal{Q}_p}$ and $f = \sum_{k=1}^n f_k g_k$. In particular, by taking $f(z) = \log(1 - ze^{-i\theta})/2$ we obtain

$$\left|\log \frac{1-ze^{-i\theta}}{2}\right| \preceq \log \frac{2}{1-|z|} \sum_{k=1}^n |g_k(z)|,$$

which implies (7.4).

On the other hand, let $(g_1, g_2, \cdots, g_n) \in M(\mathcal{Q}_p) \times M(\mathcal{Q}_p) \times \cdots \times M(\mathcal{Q}_p)$ and (7.4) hold. In order to show that $M_{(g_1,\cdots,g_n)} : \mathcal{Q}_p \times \mathcal{Q}_p \times \cdots \times \mathcal{Q}_p \to \mathcal{Q}_p$ is surjective, we must verify that for every $f \in \mathcal{Q}_p$, there are $f_1, f_2, \cdots, f_n \in \mathcal{Q}_p$ to ensure the equation: $\sum_{k=1}^n f_k g_k = f$. By the proof of Theorem 7.4.1, we see that h_k in (7.5) are non-holomorphic functions satisfying $\sum_{k=1}^n g_k h_k = 1$. However, if we can find functions $b_{j,k}$ $(j, k = 1, 2, \cdots, n)$ defined on $\mathbf{D}$ to guarantee $b_{j,k} \in \mathcal{Q}_p(\mathbf{T})$ and

$$\frac{\partial b_{j,k}(z)}{\partial \bar{z}} = f(z)h_j(z)\frac{\partial h_k(z)}{\partial \bar{z}},$$

on $\mathbf{D}$, then

$$f_j = fh_j + \sum_{k=1}^n (b_{j,k} - b_{k,j})g_k$$

just meet the requirements: $\sum_{k=1}^n f_k g_k = f$ and $f_j \in \mathcal{Q}_p$. Note that $fh_j \in \mathcal{Q}_p(\mathbf{T})$ can be figured out from the following argument. Obviously, we are required only to prove that $\partial b/\partial \bar{z} = fh$ (where $b = b_{j,k}$ and $h = h_j \partial h_k/\partial \bar{z}$) admits $\mathcal{Q}_p(\mathbf{T})$-solution. To this end, we choose a standard solution to $\partial b/\partial \bar{z} = fh$, that is,

$$b(z) = \frac{1}{\pi} \int_{\mathbf{D}} \frac{f(\zeta)h(\zeta)}{z - \zeta} dm(\zeta). \tag{7.8}$$

It is easy to see that this solution is C^2 on $\mathbf{D}$, but also continuous on $\mathbf{C}$. Certainly, we cannot help checking whether or not such a solution belongs to $\mathcal{Q}_p(\mathbf{T})$.

From the conditions $f \in \mathcal{Q}_p$ and $g_k \in M(\mathcal{Q}_p)$ as well as Theorems 7.4.2 it turns out that for the Carleson box $S(I)$,

$$\begin{aligned}
\int_{S(I)} (\cdots)_{\bar{\partial}} &= \int_{S(I)} \left|\frac{\partial b(z)}{\partial \bar{z}}\right|^2 (1 - |z|)^p dm(z) \\
&= \int_{S(I)} |f(z)h(z)|^2 (1 - |z|)^p dm(z) \\
&\preceq \sum_{k=1}^n \int_{S(I)} |f'(z)g_k(z)|^2 (1 - |z|)^p dm(z) \\
&+ \sum_{k=1}^n \int_{S(I)} |(fg_k)'(z)|^2 (1 - |z|)^p dm(z).
\end{aligned}$$

For convenience, we reformulate the Beurling transform of a function $\Phi \in L^1_{loc}(\mathbf{C})$ as

$$(T(\Phi))(z) = \text{p.v.} \int_{\mathbf{C}} \frac{\Phi(w)}{(z - w)^2} dm(w), \quad z \in \mathbf{C}.$$

If $\Phi = fh$ on $\mathbf{D}$ and $\Phi = 0$ on $\mathbf{D}^c$, then $\partial b/\partial z = (T(\Phi))(z)$ and hence for the Carleson box $S(I)$,

$$\begin{aligned}
\int_{S(I)} (\cdots)_\partial &= \int_{S(I)} \left|\frac{\partial b(z)}{\partial z}\right|^2 (1-|z|)^p dm(z) \\
&\le 2\int_{S(I)} |(T(1_{S(2I)}\Phi))(z)|^2 (1-|z|)^p dm(z) \\
&+ 2\int_{S(I)} |(T((1-1_{S(2I)}\Phi))(z)|^2 (1-|z|)^p dm(z) \\
&\le 4\int_{\mathbf{C}} |(T(1_{S(2I)}\Phi)(z)|^2 |1-|z||^p dm(z) \\
&+ 4\int_{S(I)} \left(\int_{\mathbf{D}\setminus S(2I)} \frac{|f(w)h(w)|}{|w-z|^2} dm(w)\right)^2 (1-|z|)^p dm(z) \\
&= Int_1 + Int_2.
\end{aligned}$$

Since $|1-|z||^p$ is an A_2-weight for $p \in (0,1)$ (cf. [44]) and the Beurling transform is a Calderón-Zygmund operator, it follows that

$$\begin{aligned}
Int_1 &\preceq \int_{\mathbf{C}} |T(1_{S(2I)\Phi})(z)|^2 |1-|z||^p dm(z) \\
&\preceq \int_{\mathbf{C}} |(1_{S(2I)fh})(z)|^2 |1-|z||^p dm(z) \\
&\preceq \int_{S(2I)} |f(z)h(z)|^2 (1-|z|)^p dm(z) \preceq |I|^p,
\end{aligned}$$

where the constants involved above and below may depend on the norms of the Beurling transform and the given functions f, h and g_k.

Due to $g_k \in M(\mathcal{Q}_p)$ once again, Theorem 7.4.2 implies

$$\int_{S(I)} |g_k'(z)|^2 (1-|z|)^p dm(z) \preceq \frac{|I|^p}{\log^2 \frac{2}{|I|}}.$$

Accordingly, by the Cauchy-Schwarz inequality one has

$$\int_{S(I)} |f(z)h(z)| dm(z) \preceq \sum_{k=1}^{n} \int_{S(I)} (|f'(z)g_k(z)| + |(gf_k)'z)|) dm(z) \preceq |I|,$$

that is to say, $fhdm$ is a 1-Carleson measure. This fact is applied to deduce

$$\begin{aligned}
Int_2 &\preceq \int_{S(I)} \left(\sum_{j=1}^{\infty} \int_{S(2^{j+1}I)\setminus S(2^j I)} \frac{|f(w)h(w)|}{|w-z|^2} dm(w)\right)^2 (1-|z|)^p dm(z) \\
&\preceq \int_{S(I)} \left(\sum_{j=1}^{\infty} \frac{1}{2^{2j}|I|^2} \int_{S(2^{j+1}I)} |f(w)h(w)| dm(w)\right)^2 (1-|z|)^p dm(z) \\
&\preceq |I|^p.
\end{aligned}$$

The above estimates on Int_j, $j = 1, 2$ tell us that

$$\int_{S(I)} \left|\frac{\partial b(z)}{\partial z}\right|^2 (1-|z|)^p dm(z) \preceq |I|^p,$$

and so that

$$\int_{S(I)} |\nabla b(z)|^2 (1-|z|)^p dm(z) \preceq |I|^p.$$

By Corollary 7.1.1 we see that b lies in $\mathcal{Q}_p(\mathbf{T})$. This completes the proof.

7.5 Interpolating Sequences

In order to solve the interpolation problem for $\mathcal{Q}_p \cap H^\infty$, we pause briefly to work with Khinchin's inequality.

Given finitely many complex numbers $a_1, \cdots, a_n$, consider the 2^n possible sums $\sum_{j=1}^n \pm a_j$ obtained as the plus-minus signs vary in the 2^n possible ways. For $q > 0$ let

$$\mathcal{E}\left(\left|\sum_{j=1}^n \pm a_j\right|^q\right)$$

denote the average value of $|\sum_{j=1}^n \pm a_j|^q$ over the 2^n choices of sign. The following lemma is a special case of the so-called Khinchin's inequality.

Lemma 7.5.1. *Let* $q \in (0, 2]$. *Then*

$$\mathcal{E}\left(\left|\sum_{j=1}^n \pm a_j\right|^q\right) \le \left(\sum_{j=1}^n |a_j|^2\right)^{q/2}. \tag{7.9}$$

Proof. The proof is from Garnett's book [66, p.302], but we include the proof for completeness. Let $n \in \mathbf{N}$ and Ω be the set of 2^n points $\omega = (\omega_1, \omega_2, \cdots, \omega_n)$, where $\omega_j = \pm 1$. Define the probability μ on Ω so that each point ω has probability 2^n. Also define $X(\omega) = \sum_{j=1}^n a_j \omega_j$. Then $X(\omega)$ is a more rigorous expression for $\sum \pm a_j$, and by definition

$$\mathcal{E}\left(\left|\sum_{j=1}^n \pm a_j\right|^q\right) = \frac{1}{2^n} \sum_{\omega\in\Omega} |X(\omega)|^q = \int_\Omega |X(\omega)|^q d\mu.$$

Meanwhile, let $X_j(\omega) = \omega_j$, $j = 1, 2, \cdots, n$. Then $|X_j(\omega)|^2 = 1$ and for $j \neq k$, $\mathcal{E}(X_j X_k) = 0$ since $X_j X_k$ takes each value ± 1 with probability $1/2$. This means that $\{X_1, X_2, \cdots, X_n\}$ are orthonormal in $L^2(\mu)$. Because $X = \sum_{j=1}^n a_j X_j$ and because $q \in (0, 2]$, Hölder's inequality implies

$$\left(\mathcal{E}\left(\left|\sum_{j=1}^{n}\pm a_j\right|^q\right)\right)^{1/q} \le \left(\int_{\Omega}|X(\omega)|^2 d\mu\right)^{1/2} = \left(\sum_{j=1}^{n}|a_j|^2\right)^{1/2}.$$

A sequence $\{z_n\}\subset \mathbf{D}$ is called an interpolating sequence for $\mathcal{Q}_p\cap H^\infty$ if for each bounded sequence $\{w_n\}\subset \mathbf{C}$ there exists a function $f\in\mathcal{Q}_p\cap H^\infty$ such that $f(z_n)=w_n$ for all $n\in\mathbf{N}$. With Lemma 7.5.1, we can establish the following theorem.

Theorem 7.5.1. *Let $p\in(0,1)$. Then a sequence $\{z_n\}\subset\mathbf{D}$ is an interpolating sequence for $\mathcal{Q}_p\cap H^\infty$ if and only if $\{z_n\}$ is separated, i.e.*

$$\inf_{m\neq n}\left|\frac{z_n-z_m}{1-\bar{z}_n z_m}\right|>0,$$

and at the same time $d\mu_{\{z_n\},p}=\sum_n(1-|z_n|^2)^p\delta_{z_n}$ is a p-Carleson measure.

Proof. The part of necessity combines Khinchin's inequality and a reproducing formula for $\mathcal{D}_p$, $p>0$. The reproducing formula of Rochberg and Wu [105] asserts that for $f\in\mathcal{D}_p$, one has

$$f(z)=f(0)+\int_{\mathbf{D}}f'(w)K(z,w)(1-|w|^2)^p\,dm(w),\quad z\in\mathbf{D},\tag{7.10}$$

where

$$K(z,w)=\frac{(1-z\bar{w})^{1+p}-1}{\bar{w}(1-z\bar{w})^{1+p}}.$$

Now assume that $\{z_n\}$ is an interpolating sequence for $\mathcal{Q}_p\cap H^\infty$. Then for $\epsilon_k^{(j)}=\pm1$, $j=1,\cdots,2^n$, $k=1,\cdots,n$, there are $f_j\in\mathcal{Q}_p\cap H^\infty$ such that $f_j(z_k)=\epsilon_k^{(j)}$, $k=1,\cdots,n$ and $\|f_j\|_{H^\infty}+|\!|\!|f_j|\!|\!|_{\mathcal{Q}_p}\preceq 1$. Applying (7.10) to $f_j\circ\sigma_w$ at $\sigma_w(z_k)$ we get

$$f_j(z_k)=f_j(w)+\int_{\mathbf{D}}(f_j\circ\sigma_w)'(\xi)K(\sigma_w(z_k),\xi)(1-|\xi|^2)^p\,dm(\xi).$$

Since

$$\begin{aligned}\sum_{k=1}^{n}(1-|\sigma_w(z_k)|^2)^p&=\sum_{k=1}^{n}\epsilon_k^{(j)}f_j(z_k)(1-|\sigma_w(z_k)|^2)^p\\&=f_j(w)\sum_{k=1}^{n}\epsilon_k^{(j)}(1-|\sigma_w(z_k)|^2)^p\\&\quad+\int_{\mathbf{D}}(f_j\circ\sigma_w)'(\xi)\sum_{k=1}^{n}\frac{\epsilon_k^{(j)}K(\sigma_w(z_k),\xi)}{(1-|\sigma_w(z_k)|^2)^{-p}}\frac{dm(\xi)}{(1-|\xi|^2)^{-p}}\\&=T_1+T_2,\end{aligned}$$

we may compute the expectation of both sides of this equality. Observe that by (7.9) with $q = 1$ we find

$$\mathcal{E}(T_1) \preceq \left(\sum_{k=1}^{n}(1-|\sigma_w(z_k)|^2)^{2p}\right)^{1/2} \preceq \left(\sum_{k=1}^{n}(1-|\sigma_w(z_k)|^2)^{p}\right)^{1/2}.$$

In the meantime, applying the Cauchy-Schwarz inequality, Lemma 1.4.1 and (7.9) with $q = 2$, we get

$$\begin{aligned}\mathcal{E}(T_2) &\leq \sup_j \|f \circ \sigma_w\|_{\mathcal{D}_p} \\ &\times \left(\int_{\mathbf{D}} \sum_{k=1}^{n} |K(\sigma_w(z_k),\xi)|^2(1-|\sigma_w(z_k)|^2)^{2p}(1-|\xi|^2)^p dm(\xi)\right)^{1/2} \\ &\preceq \left(\sum_{k=1}^{n}(1-|\sigma_w(z_k)|^2)^{2p}\int_{\mathbf{D}}\frac{(1-|\xi|^2)^p}{|1-\overline{\sigma_w(z_k)}\xi|^{2+2p}}dm(\xi)\right)^{1/2} \\ &\preceq \left(\sum_{k=1}^{n}(1-|\sigma_w(z_k)|^2)^{p}\right)^{1/2}.\end{aligned}$$

So, the estimates involving $\mathcal{E}(T_1)$ and $\mathcal{E}(T_2)$ indicate that the second condition of Theorem 7.5.1 holds. Since $\{z_n\}$ is also an interpolating sequence for H^∞, the first condition holds as well.

To demonstrate the part of sufficiency we suppose that $\{z_n\}$ satisfies the above assumption. By the Cauchy-Schwarz inequality we see that $\sum_n(1-|z_n|^2)\delta_{z_n}$ is a 1-Carleson measure and then by the argument in [66, p. 287] that $\{z_n\}$ is uniformly separated, namely,

$$\eta = \inf_n \prod_{m\neq n}\left|\frac{z_n - z_m}{1-\bar{z}_n z_m}\right| > 0.$$

Now, for any $\{w_n\} \in l^\infty$ let

$$\begin{aligned}f(z) &= \sum_n \epsilon_n w_n \left(\frac{1-|z_n|^2}{1-\bar{z}_n z}\right)^2 B_n(z) \\ &\times \exp\left(-\gamma \sum_{|z_m|\geq|z_n|}\left(\frac{1+\bar{z}_m z}{1-\bar{z}_m z}\right)(1-|z_m|^2)\right).\end{aligned}$$

Here and afterwards,

$$B_n(z) = \prod_{m\neq n}\frac{|z_m|}{z_m}\left(\frac{z_m - z}{1-\bar{z}_m z}\right),$$

in which $|z_m|/z_m$ is replaced by 1 if $z_m = 0$. Besides,

$$\epsilon_n = \frac{1}{B_n(z_n)} \exp\left(\gamma \sum_{|z_m|\geq|z_n|} \left(\frac{1+\bar{z}_m z_n}{1-\bar{z}_m z_n}\right)(1-|z_m|^2)\right)$$

for $\gamma = 1/(2\log(e/\eta^2))$.

It is clear that $f \in H^\infty$ and $f(z_n) = w_n$ for $n \in \mathbf{N}$. However, what we want is: $f \in \mathcal{Q}_p$. As in the proof of Theorem 7.2.1, we consider

$$F(z) = -\bar{z} \sum_n \epsilon_n w_n \left(\frac{1-|z_n|^2}{|1-\bar{z}_n z|^2}\right) \times \exp\left(-\gamma \sum_{|z_m|\geq|z_n|} \left(\frac{1+\bar{z}_m z}{1-\bar{z}_m z}\right)(1-|z_m|^2)\right),$$

then $f(\zeta) = B(\zeta)F(\zeta)$ for $\zeta \in \mathbf{T}$, where

$$B(z) = \prod_m \frac{|z_m|}{z_m}\left(\frac{z_m - z}{1-\bar{z}_m z}\right),$$

in which $|z_m|/z_m$ is replaced by 1 if $z_m = 0$. By Theorem 5.2.1 we know that $B \in \mathcal{Q}_p$. Therefore, in order to prove $f \in \mathcal{Q}_p$ we only need to show that $|\nabla F(z)|^2(1-|z|^2)^p dm(z)$ is a p-Carleson measure, due to Corollary 7.1.1. The same argument as that leading to (7.3) gives that $\{\epsilon_n\} \in l^\infty$, and so that

$$|\nabla F(z)| \preceq \|\{w_n\}\|_\infty \int_{\mathbf{D}} \frac{G(w)}{|1-\bar{w}z|^2} dm(w), \tag{7.11}$$

where

$$G(w) = \sum_m \delta_{z_m}(w), \quad w \in \mathbf{D}.$$

Thus for the Carleson box $S(I)$ one has that

$$\int_{S(I)} |G(z)|^2(1-|z|^2)^p dm(z) = \sum_{z_m \in S(I)} (1-|z_m|^2)^p,$$

and so that $|G(z)|^2(1-|z|^2)^p dm(z)$ is a p-Carleson measure. Employing (7.11) and Lemma 7.2.2 we finally obtain that $|\nabla F(z)|^2(1-|z|^2)^p dm(z)$ is a p-Carleson measure. Therefore, the proof is complete.

Notes

7.1 Lemma 7.1.1 is from Stegenga [117]. Theorem 7.1.1 gives another proof for Theorem 6.1.1. The equivalences among (i), (ii), (iii) and (iv) of Theorem 7.1.1 are from the papers of Nicolau-Xiao [95] and Xiao [136], respectively. In fact, these equivalences show that $\mathcal{Q}_p(\mathbf{T})$ consists of all Möbius bounded functions in the Sobolev space $\mathcal{L}_p^2(\mathbf{T})$ on $\mathbf{T}$, namely, $f \in \mathcal{Q}_p(\mathbf{T})$ if and only if

$$\sup_{w \in \mathbf{D}} \|f \circ \sigma_w - f(w)\|_{\mathcal{L}_p^2(\mathbf{T})} < \infty;$$

see also [136]. Here we say that a measurable function f on $\mathbf{T}$ belongs to $\mathcal{L}_p^2(\mathbf{T})$ provided

$$\|f\|_{\mathcal{L}_p^2(\mathbf{T})} = \left(\int_{\mathbf{T}} \int_{\mathbf{T}} \frac{|f(w) - f(z)|^2}{|w - z|^{2-p}} |dz||dw| \right)^{\frac{1}{2}} < \infty.$$

For some relations between $BMO(\mathbf{T})$ and the Sobolev spaces, we refer to Strichartz [120].

7.2 Lemma 7.2.2 and Theorem 7.2.1 are also from [95]. Lemma 7.2.2 has been employed by Suárez to study meromorphic functions [123].

7.3 The results in Sections 7.3 and their proofs can be found in [95]. Note that the argument for Theorem 7.3.3 does not require the predual of $\mathcal{Q}_p$. At this point, $\mathcal{Q}_p$ is different from $BMOA$. Nevertheless, it would be interesting to characterize the predual of $\mathcal{Q}_p$ for each $p \in (0, 1)$.

7.4 Theorems 7.4.2 and 7.4.3 are in Xiao [135]. Observe that only necessity of Theorem 7.4.1 is useful in the proof of Theorem 7.4.2. So, these theorems have been reasonably generalized by Andersson and Carlsson to the Q spaces over strongly pseudoconvex domains of $\mathbf{C}^n$; see [9]. However, it would be interesting to give a full description of $M(\mathcal{Q}_p)$, $p \in (0, 1)$, since the cases $p \geq 1$ have been figured out by Stegenga [116], Ortega-Fabrega [96] and Brown-Shields [36], respectively. Theorem 7.4.3 is available for the cases $p = 0$ and $p \geq 1$; see Nicolau [94] and Ortega-Fabrega [96],[97].

7.5 Concerning Theorem 7.5.1 (cf. [95]), we would like to point out that Earl's constructive solution [58] for H^∞-interpolation may be modified to prove the sufficiency part of Theorem 7.5.1. In fact, Earl's construction indicates that when $\{z_n\}$ is an interpolating sequence for $\mathcal{Q}_p \cap H^\infty$ there exist interpolating functions of the form $\kappa B(z)$, where κ is a constant and $B(z)$ is a Blaschke product. The Blaschke product $B(z)$ has simple zeros $\{\zeta_n\}$ which are hyperbolically very close to the $\{z_n\}$. It follows that $\{\zeta_n\}$ is also an interpolating sequence for $\mathcal{Q}_p \cap H^\infty$. Another proof involved in $\bar{\partial}$-techniques is presented in [95].

8. Dyadic Localization

This chapter contains a local analysis of $\mathcal{Q}_p(\mathbf{T})$ based on the dyadic portions. First of all, we give an alternate characterization of $\mathcal{Q}_p$ in terms of the square mean oscillations over successive bipartitions of arcs in $\mathbf{T}$. Next, we consider the dyadic counterpart $\mathcal{Q}_p^d(\mathbf{T})$ of $\mathcal{Q}_p(\mathbf{T})$, in particular, we show that $f \in \mathcal{Q}_p(\mathbf{T})$ if and only if (almost) all its translates belong to $\mathcal{Q}_p^d(\mathbf{T})$; conversely, functions in $\mathcal{Q}_p(\mathbf{T})$ may be obtained by averaging translates of functions in $\mathcal{Q}_p(\mathbf{T})$. Finally, as a natural application of the dyadic model of $\mathcal{Q}_p(\mathbf{T})$, we present a wavelet expansion theorem of $\mathcal{Q}_p(\mathbf{T})$.

8.1 Square Mean Oscillation

From now on, using the map: $t \to e^{2\pi i t}$, we identify $\mathbf{T}$ with the unit interval $[0, 1)$, where subintervals may wrap around 0. Meanwhile, a subarc of $\mathbf{T}$ corresponds to a subinterval of $[0, 1)$. A dyadic interval is an interval of the type: $[m2^{-n}, (m+1)2^{-n})$, $n \in \mathbf{N} \cup \{0\}$, $k = 0, 1, \cdots, 2^n - 1$. Denote by $\mathcal{I}$ the set of all dyadic subintervals of $\mathbf{T}$ (including $\mathbf{T}$ itself), and let $\mathcal{I}_n$, $n \in \mathbf{N} \cup \{0\}$ be the set of the 2^n dyadic intervals of length 2^{-n}. Similarly, if $I \subseteq \mathbf{T}$ is any interval, dyadic or not, we let $\mathcal{I}_n(I)$, $n \in \mathbf{N} \cup \{0\}$, denote the set of the 2^n subintervals of length $2^{-n}|I|$ obtained by n successive bipartition of I. Of course, $|I|$ still denotes the length of interval $I \subseteq \mathbf{T}$.

For the sake of simplicity, we rewrite, for any interval $I \subseteq \mathbf{T}$ and an $L^2(I)$ function f,

$$f(I) = f_I = \frac{1}{|I|} \int_I f(x)dx,$$

the mean of f on I, and define

$$\Phi_f(I) = \frac{1}{|I|} \int_I |f(x) - f(I)|^2 dx,$$

the square mean oscillation of f on I. Obviously, $\Phi_f(I) < \infty$ if and only if $f \in L^2(I)$; we may extend the definition to all measurable functions f on I by letting $\Phi_f(I) = \infty$ when $f \notin L^2(I)$.

Recall that $f \in BMO(\mathbf{T})$ if and only if $\sup_I \Phi_f(I) < \infty$, where the supremum is taken over all intervals in $\mathbf{T}$. Moreover, the forthcoming two identities are easily verified.

$$\frac{1}{|I|}\int_I |f(x)-a|^2dx = \Phi_f(I) + |f(I)-a|^2, \quad a \in \mathbf{C}; \tag{8.1}$$

and

$$\frac{1}{|I|^2}\int_I\int_I |f(x)-f(y)|^2\,dx\,dy = 2\Phi_f(I). \tag{8.2}$$

Furthermore, if $I \subseteq J$, then by (8.1),

$$\Phi_f(I) \le \frac{1}{|I|}\int_I |f(x)-f(J)|^2dx \le \frac{|J|}{|I|}\Phi_f(J). \tag{8.3}$$

Similarly,

$$|f(I)-f(J)|^2 \le \frac{|J|}{|I|}\Phi_f(J). \tag{8.4}$$

Given an interval $I \subseteq \mathbf{T}$ and an $L^2(I)$ function f, set

$$\Psi_{f,p}(I) = \sum_{k=0}^{\infty}\sum_{J\in\mathcal{I}_k(I)} 2^{-pk}\Phi_f(J). \tag{8.5}$$

The goal of introducing (8.5) is to investigate mean oscillation of each $\mathcal{Q}_p(\mathbf{T})$ function. To achieve this we need two preparatory lemmas.

Lemma 8.1.1. *Let $p \in (0,1)$ and $I \subseteq \mathbf{T}$ be an interval. If $f \in L^2(I)$ then*

$$\Phi_f(I) = 2^{-1}\sum_{J\in\mathcal{I}_1(I)}\Phi_f(J) + 2^{-2}\sum_{J\in\mathcal{I}_1(I)}|f(J)-f(I)|^2 \tag{8.6}$$

and

$$\Psi_{f,p}(I) \approx \sum_{J\in\mathcal{I}_1(I)}\Psi_{f,p}(J) + \sum_{J\in\mathcal{I}_1(I)}|f(J)-f(I)|^2. \tag{8.7}$$

Proof. (8.6) follows from (8.1) and

$$\begin{aligned}\Phi_f(I) &= |I|^{-1}\sum_{J\in\mathcal{I}_1(I)}\int_J |f(x)-f(I)|^2dx \\ &= 2^{-1}\sum_{J\in\mathcal{I}_1(I)}\left(\Phi_f(J) + |f(J)-f(I)|^2\right).\end{aligned}$$

Next, this and (8.5) yield, since $\mathcal{I}_k(I) = \bigcup_{J\in\mathcal{I}_1(I)}\mathcal{I}_{k-1}(J)$ for $k \in \mathbf{N}$,

$$\begin{aligned}\Psi_{f,p}(I) &= \Phi_f(I) + \sum_{k=1}^{\infty}\sum_{J\in\mathcal{I}_1(I)}\sum_{K\in\mathcal{I}_{k-1}(J)} 2^{-pk}\Phi_f(K) \\ &= \Phi_f(I) + \sum_{J\in\mathcal{I}_1(I)} 2^{-p}\Psi_{f,p}(J) \\ &\approx \sum_{J\in\mathcal{I}_1(I)}\left(\Psi_{f,p}(J) + \Phi_f(J) + |f(J)-f(I)|^2\right) \\ &\approx \sum_{J\in\mathcal{I}_1(I)}\left(\Psi_{f,p}(J) + |f(J)-f(I)|^2\right),\end{aligned}$$

which implies (8.7).

Lemma 8.1.2. *Let $p \in (0,1)$ and $I \subseteq \mathbf{T}$ be an interval. If $f \in L^2(I)$, then*

$$\Psi_{f,p}(I) \preceq \frac{1}{|I|^p} \int_I \int_I \frac{|f(x) - f(y)|^2}{|x-y|^{2-p}} dxdy. \tag{8.8}$$

Proof. By (8.5) and (8.2),

$$\begin{aligned}\Psi_{f,p}(I) &= 2^{-1} \sum_{k=0}^{\infty} \sum_{J \in \mathcal{I}_k(I)} 2^{-pk} \left(2^{-k}|I|\right)^{-2} \int_J \int_J |f(x) - f(y)|^2 \, dx \, dy \\ &= \int_{\mathbf{T}} \int_{\mathbf{T}} \alpha_I(x,y) |f(x) - f(y)|^2 \, dx \, dy. \end{aligned} \tag{8.9}$$

Here and afterwards,

$$\alpha_I(x,y) = 2^{-1} \sum_{k=0}^{\infty} \sum_{J \in \mathcal{I}_k(I)} 2^{(2-p)k} |I|^{-2} 1_J(x) 1_J(y). \tag{8.10}$$

Since $x, y \in J \in \mathcal{I}_k(I)$ implies $|x-y| \le 2^{-k}|I|$, one has

$$\alpha_I(x,y) \le \sum_{2^k \le |I|/|x-y|} 2^{(2-p)k} |I|^{-2} \preceq |x-y|^{p-2} |I|^{-p};$$

furthermore $\alpha_I(x,y) = 0$ unless $x, y \in I$. Consequently, (8.8) follows.

Although we will show the converse inequality of (8.8), before doing so we define $I + t = \{x + t; x \in I\}$ for I (an interval in $\mathbf{T}$) and $t \in \mathbf{R}$, and then give a slightly weaker but more general form.

Lemma 8.1.3. *Let $p \in (0,1)$ and $I \subseteq \mathbf{T}$ be an interval. If $f \in L^2(I)$, then*

$$\frac{1}{|I|^p} \int_I \int_I \frac{|f(x) - f(y)|^2}{|x-y|^{2-p}} dxdy \preceq \frac{1}{|I|} \int_{-|I|}^{|I|} \Psi_{f,p}(I+t) \, dt + \Psi_{f,p}(I).$$

Proof. By (8.9) and Fubini's theorem,

$$\frac{1}{|I|} \int_{-|I|}^{|I|} \Psi_{f,p}(I+t) \, dt = \int_{\mathbf{T}} \left(\int_{\mathbf{T}} \frac{1}{|I|} \int_{-|I|}^{|I|} \alpha_{I+t}(x,y) \, dt \right) |f(x) - f(y)|^2 \, dx \, dy.$$

This and (8.9) show that it suffices to verify

$$\frac{1}{|I|} \int_{-|I|}^{|I|} \alpha_{I+t}(x,y) \, dt + \alpha_I(x,y) \succeq \frac{|x-y|^{p-2}}{|I|^p}, \qquad x, y \in I. \tag{8.11}$$

First, suppose that $x, y \in I$ with $|x - y| \le |I|/2$ and let $l \in \mathbf{N} \cup \{0\}$ be such that $2^{-l-2}|I| < |x-y| \le 2^{-l-1}|I|$. Then, by (8.10), and noting that $x \notin I + t$ and thus $\alpha_{I+t}(x,y) = 0$ when $|t| > |I|$,

$$\begin{aligned}\frac{1}{|I|}\int_{-|I|}^{|I|} \alpha_{I+t}(x,y)\,dt &\geq \frac{1}{2|I|}\int_{\mathbf{T}} \sum_{J\in\mathcal{I}_l(I+t)} 2^{(2-p)l}|I|^{-2}1_J(x)1_J(y)\,dt \\ &= \frac{2^{(2-p)l}}{4|I|^3}\sum_{J\in\mathcal{I}_l(I)}\int_{\mathbf{T}} 1_{J+t}(x)1_{J+t}(y)\,dt \\ &\succeq \frac{|x-y|^{p-2}}{|I|^{p+1}}\sum_{J\in\mathcal{I}_l(I)}\int_{\mathbf{T}} 1_{J+t}(x)1_{J+t}(y)\,dt.\end{aligned}$$

It is easily seen that the final integral, for each J, equals $|J|-|x-y|\geq |I|/2$, and thus the sum over J is at least $|J|/2$, and (8.11) holds for $|x-y|\leq |I|/2$.

Finally, if $x,y\in I$ with $|x-y|>|I|/2$, then, by (8.10),

$$\alpha_I(x,y)\geq 2^{-1}|I|^{-2}\succeq |I|^{-p}|x-y|^{p-2}$$

and (8.11) holds in this case too.

To produce a full converse to the inequality in Lemma 8.1.2, we still need two more lemmas, which may also have independent interest.

Lemma 8.1.4. *Let $p\in(0,1)$. Let $I,I',I''\subseteq\mathbf{T}$ be three intervals of equal size: $|I|=|I'|=|I''|$, such that I' and I'' are adjacent and $I\subseteq I'\cup I''$. Then, for any $f\in L^2(I'\cup I'')$,*

$$\Phi_f(I)\leq \Phi_f(I')+\Phi_f(I'')+|f(I')-f(I'')|^2, \tag{8.12}$$

and

$$\Psi_{f,p}(I)\preceq \Psi_{f,p}(I')+\Psi_{f,p}(I'')+|f(I')-f(I'')|^2. \tag{8.13}$$

Proof. It follows from (8.3) and (8.6) that

$$\Phi_f(I)\leq \frac{|I'\cup I''|}{|I|}\Phi_f(I'\cup I'')=\Phi_f(I')+\Phi_f(I'')+\frac{1}{2}|f(I')-f(I'')|^2,$$

proving (8.12).

Regarding (8.13), we assume for simplicity that $I'=[0,1)$ and $I''=[1,2)$; this is no loss of generality by homogeneity. For each integer $j\geq 0$, let $\{I_{j,k}\}_{k=1}^{2^{j+1}}$ be the set of the 2^{j+1} dyadic intervals of length 2^{-j} contained in $I'\cup I''$ (i.e., $\mathcal{I}_j(I')\cup\mathcal{I}_j(I'')$), arranged in the natural order. If $J\in\mathcal{I}_j(I)$, then $J\subseteq I_{j,k}\cup I_{j,k+1}$ for some k, and thus by (8.12) applied to J,

$$\Phi_f(I)\leq \Phi_f(I_{j,k})+\Phi_f(I_{j,k+1})+|f(I_{j,k})-f(I_{j,k+1})|^2.$$

The 2^j different choices of $J\in\mathcal{I}_j(I)$ yield different k, and summing over all j and J we thus obtain,

$$\begin{aligned}\Psi_{f,p}(I) &= \sum_{j=0}^{\infty} \sum_{J \in \mathcal{I}_j(I)} 2^{-pj} \Phi_f(J) \\ &\leq 2 \sum_{j=0}^{\infty} \sum_{k=1}^{2^{j+1}} \frac{\Phi_f(I_{j,k})}{2^{pj}} + \sum_{j=0}^{\infty} \sum_{k=1}^{2^{j+1}-1} \frac{|f(I_{j,k}) - f(I_{j,k+1})|^2}{2^{pj}}. \qquad (8.14)\end{aligned}$$

The first double sum on the right hand side of the last inequality is just $\Psi_{f,p}(I') + \Psi_{f,p}(I'')$. In order to dominate the final double sum, consider a pair (j,k) with $j \geq 0$ and $1 \leq k < 2^{j+1} - 1$. Let I^* be the smallest dyadic interval that contains $I_{j,k} \cup I_{j,k+1}$, and let the length of I^* be 2^{-j+m}, where $m \in \mathbf{N}$. Moreover, for $0 \leq l \leq m$, let J_l and K_l be the dyadic intervals of length 2^{-j+l} that contain $I_{j,k}$ and $I_{j,k+1}$, respectively; thus $I_{j,k} = J_0 \subset J_1 \subset \cdots \subset J_m = I^*$ and $I_{j,k+1} = K_0 \subset \cdots \subset K_m = I^*$. Using the Cauchy-Schwarz inequality and (8.4), we get

$$\begin{aligned}|f(I_{j,k}) - f(I_{j,k+1})|^2 &\leq \Big(\sum_{l=1}^{m} |f(J_{l-1}) - f(J_l)| + \sum_{l=1}^{m} |f(K_l) - f(K_{l-1})|\Big)^2 \\ &\preceq \sum_{l=1}^{m} l^2 |f(J_l) - f(J_{l-1})|^2 + \sum_{l=1}^{m} l^2 |f(K_l) - f(K_{l-1})|^2 \\ &\preceq \sum_{l=1}^{m} l^2 (\Phi_f(J_l) + \Phi_f(K_l)). \qquad (8.15)\end{aligned}$$

If $k \neq 2^j$, then $I_{j,k} \cup I_{j,k+1} \subseteq I'$ or I'', and thus $|I^*| \leq 1$ and $m \leq j$. If $k = 2^j$ however, then $I^* = [0,2)$ and $m = j+1$; in this case we modify (8.15) by observing that $J_j = I'$ and $K_j = I''$ and thus

$$\begin{aligned}|f(I_{j,k}) - f(I_{j,k+1})| \leq \sum_{l=1}^{m} |f(J_{l-1}) - f(J_l)| + \sum_{l=1}^{m} |f(K_l) - f(K_{l-1})| \\ + |f(I') - f(I'')|\end{aligned}$$

which by the same argument implies

$$|f(I_{j,k}) - f(I_{j,k+1})|^2 \preceq \sum_{l=1}^{m} l^2 (\Phi_f(J_l) + \Phi_f(K_l)) + |f(I') - f(I'')|^2. \qquad (8.16)$$

We now keep $j \geq 0$ fixed and sum (8.15) or (8.16) (when $k = 2^j$) for $1 \leq k \leq 2^{j+1} - 1$. We observe that the intervals J_l and K_l that appear belong to $\mathcal{I}_{j-l}(I') \cup \mathcal{I}_{j-l}(I'')$, with $1 \leq l \leq j$. Moreover, each dyadic interval J in $\mathcal{I}_{j-l}(I') \cup \mathcal{I}_{j-l}(I'')$ appears at most four times as a J_l or a K_l (viz. when, in $\mathcal{I}_l(J)$, $I_{j,k}$ is the rightmost interval, $I_{j,k+1}$ is the leftmost interval or $I_{j,k}$ and $I_{j,k+1}$ are the two middle intervals). By noting that (8.16) is used only once, we then have

$$\sum_{k=1}^{2^{j+1}-1} |f(I_{j,k}) - f(I_{j,k+1})|^2 \preceq \sum_{l=1}^{j} \sum_{J\in\mathcal{I}_{j-l}(I')\cup\mathcal{I}_{j-1}(I'')} l^2\Phi_f(J) + |f(I') - f(I'')|^2.$$

Substituting $j = k + l$ and summing over j, we finally obtain

$$\begin{aligned}\sum_{j=0}^{\infty}\sum_{k=1}^{2^{j+1}-1} 2^{-pj}|f(I_{j,k}) - f(I_{j,k+1})|^2 &\preceq \sum_{l=1}^{\infty}\sum_{k=0}^{\infty}\sum_{J\in\mathcal{I}_k(I')\cup\mathcal{I}_k(I'')} 2^{-pk-pl}l^2\Phi_f(J)\\ &\quad+\sum_{j=0}^{\infty}2^{-pj}|f(I') - f(I'')|^2\\ &\preceq \Psi_{f,p}(I') + \Psi_{f,p}(I'') + |f(I') - f(I'')|^2,\end{aligned}$$

which completes the proof of (8.13) via (8.14).

Theorem 8.1.1. *Let $p \in (0,1)$ and $f \in L^2(\mathbf{T})$. Then $f \in \mathcal{Q}_p(\mathbf{T})$ if and only if $\sup_{I\subseteq\mathbf{T}} \Psi_{f,p}(I)$ is finite, where I ranges over all intervals in $\mathbf{T}$. Moreover, for any interval $I \subseteq \mathbf{T}$ and $f \in L^2(I)$,*

$$\Psi_{f,p}(I) \approx \frac{1}{|I|^p}\int_I\int_I \frac{|f(x) - f(y)|^2}{|x-y|^{2-p}}dxdy.$$

Proof. Necessity follows immediately from Lemma 8.1.2. For sufficiency, we may assume that f is defined on the real line $\mathbf{R}$ with constant $f(I)$ outside I. Let I_- and I_+ be the two intervals of the same length as I that are adjacent to I on the left and the right, respectively. Note that then $\Psi_{f,p}(I_-) = \Psi_{f,p}(I_+) = 0$ and that $f(I_-) = f(I_+) = f(I)$.

For each t with $|t| < |I|$, either $I + t \subset I_- \cup I$ or $I + t \subset I_+ \cup I$, and in both cases Lemma 8.1.4 shows $\Psi_{f,p}(I + t) \preceq \Psi_{f,p}(I)$. The result follows by Lemma 8.1.3.

Corollary 8.1.1. *Let $p \in (0,1)$. Then $\mathcal{Q}_p(\mathbf{T})$ equals the space of all $f \in L^2(\mathbf{T})$ such that $\sup_I \Psi_{f,p}(I)$ is finite, where I ranges over all intervals in $\mathbf{T}$ with dyadic length 2^{-n}, $n \in \mathbf{N}\cup\{0\}$*

Proof. Since every interval I is contained in an interval J with dyadic length $|J| < 2|I|$, it is obvious that it suffices to consider intervals of dyadic lengths in the definition of $\mathcal{Q}_p(\mathbf{T})$. The proof is completed by Theorem 8.1.1.

8.2 Dyadic Model

Motivated by Corollary 8.1.1, we define a dyadic analogue of $\mathcal{Q}_p(\mathbf{T})$ and give some results involving the two spaces.

First, the dyadic distance $d(x, y)$ between two points in $\mathbf{T}$ is determined by

$$d(x,y) = \inf\{|I| : \; x, y \in I \in \mathcal{I}\}.$$

For $p \in (0,1)$, the space $\mathcal{Q}_p^d(\mathbf{T})$ is defined as the class of all functions $f \in L^2(\mathbf{T})$ such that

$$\|f\|^2_{\mathcal{Q}_p^d} = \sup_{I\in\mathcal{I}} \frac{1}{|I|^p} \int_I \int_I \frac{|f(x)-f(y)|^2}{(d(x,y))^{2-p}} dxdy < \infty.$$

Meanwhile, $BMO^d(\mathbf{T})$ stands for the dyadic version of $BMO(\mathbf{T})$, namely, $f \in BMO^d(\mathbf{T})$ provided f is an $L^2(\mathbf{T})$ function satisfying

$$\|f\|^2_{BMO^d} = \sup_{I\in\mathcal{I}} \frac{1}{|I|^2} \int_I \int_I |f(x)-f(y)|^2 dxdy < \infty.$$

Since $d(x,y) \geq |x-y|$, it follows immediately that $\mathcal{Q}_p(\mathbf{T}) \subseteq \mathcal{Q}_p^d(\mathbf{T})$. The inclusion is strict; for example, it is easily seen that if $f(x) = \log x$, $x \in (0,1)$, then $f \in \mathcal{Q}_p^d(\mathbf{T})$, but $f \notin BMO(\mathbf{T})$ because of the infinite jump at 0, and thus $f \notin \mathcal{Q}_p(\mathbf{T})$.

Theorem 8.2.1. *Let $p \in (0,1)$ and $f \in L^2(\mathbf{T})$. Then $f \in \mathcal{Q}_p^d(\mathbf{T})$ if and only if $\sup_{I\in\mathcal{I}} \Psi_{f,p}(I) < \infty$.*

Proof. Necessity follows immediately from the definition. For sufficiency, we observe that if $I \in \mathcal{I}$, $x, y \in I$ and $k \in \mathbf{N} \cup \{0\}$, then $x, y \in J$ for some $J \in \mathcal{I}_k(I)$ if and only if $d(x,y) \leq 2^{-k}|I|$, and thus, by (8.10),

$$\alpha_I(x,y) = 2^{-1} \sum_{2^k \leq |I|/d(x,y)} 2^{(2-p)k}|I|^{-2} \approx |I|^{-p}(d(x,y))^{p-2}.$$

The desired result follows from (8.9).

For $n \in \mathbf{N} \cup \{0\}$ let $\mathcal{F}_n$ be the σ-field generated by the partition $\mathcal{I}_n$; and for $f \in L^2(\mathbf{T})$ let $E(f|\mathcal{F}_n)$ stand for the function that is constant $f(I)$ on each dyadic interval $I \in \mathcal{I}_n$.

Theorem 8.2.2. *Let $p \in (0,1)$. If $f \in L^2(\mathbf{T})$ and $f_n = E(f|\mathcal{F}_n)$, then the following conditions are equivalent:*

(i) $f \in \mathcal{Q}_p^d(\mathbf{T})$.

(ii) *For $n \in \mathbf{N} \cup \{0\}$,*

$$\sum_{k=0}^{\infty} 2^{(1-p)k} E(|f - f_{n+k}|^2|\mathcal{F}_n) \preceq 1, \quad a.s..$$

(iii) *For $n \in \mathbf{N} \cup \{0\}$,*

$$\sum_{k=0}^{\infty} 2^{(1-p)k} E(|f_{n+k+1} - f_{n+k}|^2|\mathcal{F}_n) \preceq 1, \quad a.s..$$

Proof. If $I \in \mathcal{I}_n$ and $J \in \mathcal{I}_k(I) \subseteq \mathcal{I}_{n+k}$, then

$$\Phi_f(J) = |J|^{-1} \int_J |f(x) - f_{n+k}|^2 dx,$$

and $|J| = 2^{-k}|I|$. Hence, by definition,

$$\Psi_{f,p}(I) = \sum_{k=0}^{\infty} \frac{2^{(1-p)k}}{|I|} \int_I |f(x) - f_{n+k}|^2 dx = \sum_{k=0}^{\infty} 2^{(1-p)k} E(|f - f_{n+k}|^2 | \mathcal{F}_n),$$

which together with Theorem 8.2.1 shows the equivalence $(i) \iff (ii)$.

Furthermore,

$$E(|f - f_{n+k}|^2 | \mathcal{F}_n) = \sum_{j=k}^{\infty} E(|f_{n+j+1} - f_{n+j}|^2 | \mathcal{F}_n),$$

and thus, interchanging the order of summation,

$$\begin{aligned}\sum_{k=0}^{\infty} 2^{(1-p)k} E(|f - f_{n+k}|^2 | \mathcal{F}_n) &= \sum_{j=0}^{\infty} \sum_{k=0}^{j} 2^{(1-p)k} E(|f_{n+j+1} - f_{n+j}|^2 | \mathcal{F}_n) \\ &\approx \sum_{j=0}^{\infty} 2^{(1-p)j} E(|f_{n+j+1} - f_{n+j}|^2 | \mathcal{F}_n),\end{aligned}$$

which implies that (ii) is equivalence to (iii).

The next theorem gives a surprising relation linking $\mathcal{Q}_p(\mathbf{T})$, $BMO(\mathbf{T})$ and $\mathcal{Q}_p^d(\mathbf{T})$.

Theorem 8.2.3. *Let $p \in (0,1)$. Then $\mathcal{Q}_p(\mathbf{T}) = \mathcal{Q}_p^d(\mathbf{T}) \cap BMO(\mathbf{T})$.*

Proof. The inclusion $\mathcal{Q}_p(\mathbf{T}) \subseteq \mathcal{Q}_p^d(\mathbf{T}) \cap \mathrm{BMO}(\mathbf{T})$ follows directly from the related definitions.

Conversely, suppose that $f \in \mathcal{Q}_p^d(\mathbf{T}) \cap \mathrm{BMO}(\mathbf{T})$. Let $I \subseteq \mathbf{T}$ be an interval of dyadic length. Then there exist two adjacent dyadic intervals I' and I'' of the same size as I, such that $I \subset I' \cup I''$. Lemma 8.1.4, Theorem 8.2.1 and (8.4) produce

$$\Psi_{f,p}(I) \preceq \Psi_{f,p}(I') + \Psi_{f,p}(I'') + |f(I') - f(I'')|^2 \preceq \|f\|^2_{\mathcal{Q}_p^d} + \|f\|^2_{BMO}.$$

Hence $\Psi_{f,p}(I)$ is bounded uniformly for all intervals I of dyadic length, and the result follows from Corollary 8.1.1.

It is clear that one reason for the discrepancy between $\mathcal{Q}_p(\mathbf{T})$ and $\mathcal{Q}_p^d(\mathbf{T})$ is that $\mathcal{Q}_p(\mathbf{T})$ is translation (i.e. rotation) invariant while $\mathcal{Q}_p^d(\mathbf{T})$ is not. In fact, the forthcoming theorem shows that a function belongs to $\mathcal{Q}_p(\mathbf{T})$ if and only if all its translates lie in $\mathcal{Q}_p^d(\mathbf{T})$. To be more precise, we denote by τ_t the translation operator: $(\tau_t f)(x) = f(x - t)$. With this notation, we have the following conclusion.

Theorem 8.2.4. *Let $p \in (0,1)$ and $f \in L^2(\mathbf{T})$. Then $f \in \mathcal{Q}_p^d(\mathbf{T})$ if and only if $\sup_{t\in\mathbf{T}}\|\tau_t f\|_{\mathcal{Q}_p^d} < \infty$.*

Proof. Since every interval of dyadic length is the translate of a dyadic interval, Corollary 8.1.1 and Theorem 8.2.1 prove that

$$\|f\|^2_{\mathcal{Q}_{p,*}} \approx \sup_{t\in\mathbf{T}} \sup_{I\in\mathcal{I}} \Psi_{f,p}(I-t) = \sup_{t\in\mathbf{T}} \sup_{I\in\mathcal{I}} \Psi_{\tau_t f}(I) \approx \sup_{t\in\mathbf{T}} \|\tau_t f\|^2_{\mathcal{Q}_p^d},$$

and so that the result follows.

The condition that $\tau_t f \in \mathcal{Q}_p^d(\mathbf{T})$ for all $t \in \mathbf{T}$ may be relaxed considerably. To see this, we need two auxiliary lemmas.

Lemma 8.2.1. *Let $I \subseteq \mathbf{T}$ be an interval of length 2^{-n}, $n \in \mathbf{N}$, and let $m(t)$, for $t \in \mathbf{T}$, be the smallest integer such that the translated interval $I + t$ is contained in a dyadic interval of length $2^{-n+m(t)}$. Then*

$$|\{t : m(t) > M\}| \le 2^{-M}, \quad M \in \mathbf{N} \tag{8.17}$$

Especially, for $r \in (0,\infty)$,

$$\int_{\mathbf{T}} (m(t))^r dt \le \sum_{k=1}^{\infty} k^r 2^{1-k} < \infty.$$

Proof. Since $|J| = 2^{-n+m(t)} \le 1$ by the definition of $m(t)$, $m(t) \le n$, and hence (8.17) is trivial for $M \ge n$. For the integer $M : 0 \le M < n$, let $I = [a, a+2^{-n})$, $a \in [0, 1-2^{-n})$. Note that $n > m(t) > M$ is equivalent to $I + t \not\subseteq J$ for any dyadic interval J with $|J| = 2^{-n+M}$. While the latter is just to say $t \in [k2^{-n+M} - 2^{-n} - a,\ k2^{-n+M} - a)$ for $k = 1, 2, ..., 2^{n-M}$. So, $\{t : m(t) > M\}$ consists of 2^{n-M} intervals of length 2^{-n}, and thus there is equality in (8.17). In particular, $|\{t : m(t) = M\}| \le 2^{1-M}$, and

$$\int_{\mathbf{T}} (m(t))^r dt = \sum_{M=0}^{n} M^r |\{t : m(t) = M\}| \le \sum_{k=1}^{\infty} k^r 2^{1-k} < \infty.$$

We are done.

Lemma 8.2.2. *Let $E \subseteq \mathbf{T}$ be a set with positive measure. If $\tau_t f \in BMO^d(\mathbf{T})$ for $t \in E$, then $f \in BMO(\mathbf{T})$.*

Proof. Let $M \in \mathbf{N}$ be such that $2^{-M} < |E|$. Since $\|\tau_t f\|_{BMO^d}$ is a measurable function in t, there exists a positive constant $C < \infty$ and a subset $E_0 \subseteq E$ with $|E_0| > 2^{-M}$ such that $\|\tau_t f\|_{BMO^d} \le C$ for $t \in E_0$.

Suppose that $I \subseteq \mathbf{T}$ is an interval of dyadic length 2^{-n} with $n \ge M$, and let $m(t)$ be in Lemma 8.2.1. By (8.17) and our assumptions,

$$|\{t : m(t) > M\}| \le 2^{-M} < |E_0|,$$

and thus there is a $t \in E_0$ such that $m(t) \le M$. Then $\|\tau_t f\|_{BMO^d} \le C$ and $I+t$ is contained in a dyadic interval J with $|J|/|I| = 2^{m(t)} \le 2^M$. Hence it follows from (8.3) that

$$\Phi_f(I) = \Phi_{\tau_t f}(I+t) \le \frac{|J|}{|I|}\Phi_{\tau_t f}(J) \le 2^M \|\tau_t f\|^2_{BMO^d} \le 2^M C^2.$$

Consequently, $\Phi_f(I)$ is uniformly bounded for all intervals $I \subseteq \mathbf{T}$ of dyadic length $\le 2^{-M}$; this easily implies, by (8.3) and (8.6), that $\Phi_f(I)$ is uniformly bounded for all intervals $I \subseteq \mathbf{T}$, i.e., $f \in BMO(\mathbf{T})$.

Theorem 8.2.5. *Let $p \in (0,1)$ and $f \in L^2(\mathbf{T})$. Then the following conditions are equivalent:*

(i) $f \in \mathcal{Q}_p(\mathbf{T})$.
(ii) $\tau_t f \in \mathcal{Q}_p^d(\mathbf{T})$ *for all* $t \in \mathbf{T}$.
(iii) $\tau_t f \in \mathcal{Q}_p^d(\mathbf{T})$ *for almost all* $t \in \mathbf{T}$.
(iv) $\tau_t f \in \mathcal{Q}_p^d(\mathbf{T})$ *for* $t \in E \subseteq \mathbf{T}$ *with* $|E| > 0$.

Proof. By Theorem 8.2.4, it remains only to show that (*iv*) implies (*i*). If (*iv*) holds, then $\tau_t f \in BMO^d(\mathbf{T})$ for $t \in E \subseteq \mathbf{T}$ with $|E| > 0$ since $\mathcal{Q}_p^d(\mathbf{T}) \subseteq BMO^d(\mathbf{T})$, and hence Lemma 8.2.2 gives $f \in BMO(\mathbf{T})$.

Now, choose some $t \in E$. Since $BMO(\mathbf{T})$ is translation invariant, $\tau_t f$ is in $BMO(\mathbf{T})$; furthermore, $\tau_t f$ belongs to $\mathcal{Q}_p^d(\mathbf{T})$ by assumption. Hence $\tau_t f \in \mathcal{Q}_p(\mathbf{T})$ by Theorem 8.2.3, and finally $f \in \mathcal{Q}_p(\mathbf{T})$, namely, (*i*) holds, due to the translation invariance of $\mathcal{Q}_p(\mathbf{T})$.

Corollary 8.2.1. *Let $p \in (0,1)$ and $f \in L^2(\mathbf{T})$. If $f \in \mathcal{Q}_p(\mathbf{T})$ then $\tau_t f \in \mathcal{Q}_p^d(\mathbf{T})$ for all $t \in \mathbf{T}$, while if $f \notin \mathcal{Q}_p(\mathbf{T})$ then $\tau_t f \notin \mathcal{Q}_p^d(\mathbf{T})$ for a.e. $t \in \mathbf{T}$.*

For $q \in (0,\infty]$, define $L^q(\mathcal{Q}_p^d)$ as the class of all measurable functions F on $\mathbf{T}\times\mathbf{T}$, such that $F(t,\cdot)$ is in $\mathcal{Q}_p^d(\mathbf{T})$ for a.e. $t \in \mathbf{T}$ and $\|F(t,\cdot)\|_{\mathcal{Q}_p^d}$ (as a function of t) belongs to $L^q(\mathbf{T})$. As the second immediate consequence of Theorem 8.2.5, we have

Corollary 8.2.2. *Let $p \in (0,1)$ and $q \in (0,\infty]$. Then $f \in \mathcal{Q}_p(\mathbf{T})$ if and only if $\tau_{(\cdot)} f(\cdot) \in L^q(\mathcal{Q}_p^d)$.*

On the other hand, if $q > 1$, beginning with any function in $L^q(\mathcal{Q}_p^d)$, we may construct a function in $\mathcal{Q}_p(\mathbf{T})$ as a suitable average. That is to say,

Theorem 8.2.6. *Let $p \in (0,1)$ and $q \in (1,\infty]$. If $F(t,\cdot) \in L^q(\mathcal{Q}_p^d)$ and $g(x) = \int_{\mathbf{T}} F(t, x+t)dt$, then $g \in \mathcal{Q}_p(\mathbf{T})$.*

Proof. Write $f_t(x) = F(t,x)$ and $h_t(x) = F(t, x+t)$. Assume that $I \subseteq \mathbf{T}$ is an interval of dyadic length 2^{-n}, $n \in \mathbf{N}$. Fix $t \in \mathbf{T}$ and let (ignoring the case when $I+t$ is dyadic) I' and I'' be the two dyadic intervals of length 2^{-n} that intersect I. Then $I \subseteq I' \cup I''$ and Lemma 8.1.4 gives

$$\begin{aligned}\Psi_{h_t,p}(I) &= \Psi_{f_t,p}(I+t)\\ &\preceq \Psi_{f_t,p}(I') + \Psi_{f_t,p}(I'') + |f_t(I') - f_t(I'')|^2\\ &\preceq \|f_t\|^2_{\mathcal{Q}_p^d} + |f_t(I') - f_t(I'')|^2.\end{aligned}$$

Let $m(t)$ be as in Lemma 8.2.1, and let, for $l = 0, ..., m(t)$, J_l and K_l be the dyadic intervals of length 2^{-n+l} that contain I' and I'', respectively. Then $J_{m(t)} = K_{m(t)}$ and, using (8.4) we get

$$\begin{aligned}|f_t(I') - f_t(I'')| &\le \sum_{l=1}^{m(t)} \left(|f_t(J_{l-1}) - f(J_l)| + |f_t(K_{l-1}) - f(K_l)|\right)\\ &\preceq \sum_{l=1}^{m(t)} \left((\Phi_{f_t}(J_l))^{1/2} + (\Phi_{f_t}(K_l))^{1/2}\right)\\ &\preceq 2m(t)\|f_t\|_{\mathcal{Q}_p^d}.\end{aligned}$$

As a consequence, we moreover obtain

$$\Psi_{h_t,p}(I) \preceq (m(t)\|f_t\|_{\mathcal{Q}_p^d})^2.$$

Because $g = \int_{\mathbf{T}} h_t dt$ and $(\Psi_{f,p}(I))^{1/2}$ may be regarded as an L^2-norm, we may use Minkowski's inequality to obtain

$$(\Psi_{g,p}(I))^{1/2} \le \int_{\mathbf{T}} (\Psi_{h_t,p}(I))^{1/2} dt \preceq \int_{\mathbf{T}} m(t)\|f_t\|_{\mathcal{Q}_p^d} dt.$$

So, it follows from Hölder's inequality and choosing $r \in (1, \infty)$ with $1/r + 1/q = 1$ that

$$(\Psi_{g,p}(I))^{1/2} \preceq \|m(t)\|_{L^r(\mathbf{T})} \|F\|_{L^q(\mathcal{Q}_p^d)}.$$

By Lemma 8.2.1, this shows that $\Psi_{g,p}(I)$ is uniformly bounded when $I \subseteq \mathbf{T}$ is an interval of dyadic length $\le 1/2$. The case $I = \mathbf{T}$ (which may be written as $[0, 2^{-1}) \cup [2^{-1}, 1)$) follows easily, and thus $g \in \mathcal{Q}_p(\mathbf{T})$ by Corollary 8.1.1.

Once we introduce the linear operator: $(Tf)(t, x) = f(x - t)$ (mapping functions on $\mathbf{T}$ to functions on $\mathbf{T} \times \mathbf{T}$), we have the following result.

Corollary 8.2.3. *Let $p \in (0,1)$ and $q \in (1, \infty]$. Then*

(i) *$\mathcal{Q}_p(\mathbf{T})$ is isomorphic to the complemented subspace $T(\mathcal{Q}_p(\mathbf{T}))$ of $L^q(\mathcal{Q}_p^d)$.*

(ii) *The adjoint operator T^* of T maps $L^q(\mathcal{Q}_p^d)$ onto $\mathcal{Q}_p(\mathbf{T})$.*

Proof. Since T^* is given by

$$T^* F(x) = \int_{\mathbf{T}} F(t, x + t) dt,$$

T^*T is the identity and TT^* is a projection. This, together with Theorem 8.2.6 implies Corollary 8.2.3.

8.3 Wavelets

The purpose of this section is to show that the well known characterization of $BMO(\mathbf{T})$ by means of a periodic wavelet basis can be extended to $\mathcal{Q}_p(\mathbf{T})$.

We start with recalling the Haar system on $\mathbf{T}$. In this section, let H denote the Haar function:

$$H(t) = \begin{cases} 1, & t \in [0,1/2), \\ -1, & t \in [1/2,1), \\ 0, & \text{otherwise}. \end{cases}$$

For $j \in \mathbf{N} \cup \{0\}$ and $k = 0, 1, \cdots, 2^j - 1$ and define $h_{j,k}(t) = 2^{j/2} H(2^j t - k)\big|_{[0,1)}$. Set also $h_{0,0}(t) = 1$. The system $\{h_{j,k}\}$ is called the Haar system on $\mathbf{T}$, and forms a complete orthonormal basis in $L^2(\mathbf{T})$. More precisely (cf. [65]), if $I_{j,k} = [k2^{-j}, (k+1)2^{-j})$ is a dyadic interval in $[0,1)$, and if

$$f_j(x) = \sum_{k=0}^{2^j-1} f(I_{j,k}) 1_{I_{j,k}}(x),$$

is an approximation of f at the resolution 2^{-j}, then it follows immediately from the Lebesgue differentiation theorem that $\lim_{j\to\infty} f_j(x) = f(x)$ a.e. on $[0,1)$. Thus, for almost every $x \in [0,1)$,

$$f(x) = f_0(x) + \sum_{j=0}^{\infty} \Big(f_{j+1}(x) - f_j(x) \Big).$$

However a simple argument shows

$$f_{j+1}(x) - f_j(x) = \sum_{k=0}^{2^j-1} \langle f, h_{j,k} \rangle h_{j,k}(x),$$

where $\langle f, g \rangle$ means the usual inner product $\int_{\mathbf{T}} f(x)\overline{g(x)}dx$. Therefore,

$$f = f(\mathbf{T}) + \sum_{j=1}^{\infty} \sum_{k=0}^{2^j-1} \langle f, h_{j,k} \rangle h_{j,k},$$

which shows that the Haar system represents f a.e. on $\mathbf{T}$, as well as in $L^2(\mathbf{T})$.

In what follows, as to $\lambda = (j,k)$ we write the shorter notation $h_{j,k}$ as h_λ and denote by $I(\lambda)$ the dyadic interval $\{t : 2^j t - k \in [0,1)\}$. Moreover, for $I \in \mathcal{I}$, a sequence $\mathbf{a} = \{a(\lambda)\}$, and $q \in (0,1]$, let

$$T_{\mathbf{a},q}(I) = |I|^{-1} \sum_{I(\lambda) \subseteq I} \left(\frac{|I|}{|I(\lambda)|} \right)^{1-q} |a(\lambda)|^2.$$

Lemma 8.3.1. *Let $p \in (0,1)$. Then for each $I \in \mathcal{I}$ and sequence $\mathbf{a} = \{\mathbf{a}(\lambda)\}$,*

$$T_{\mathbf{a},p}(I) \approx \sum_{k=0}^{\infty} 2^{-pk} \sum_{J \in \mathcal{I}_k(I)} T_{\mathbf{a},1}(J).$$

Proof. The right hand side equals

$$\sum_{J \in \mathcal{I}_1(I)} \sum_{I(\lambda) \subseteq J} \left(\frac{|J|}{|I|}\right)^p \frac{|a(\lambda)|^2}{|J|} = |I|^{-p} \sum_{I(\lambda) \subseteq I} |a(\lambda)|^2 \sum_{J \in \mathcal{I}_1(I), J \supseteq I(\lambda)} |J|^{p-1}$$
$$\approx |I|^{-p} \sum_{I(\lambda) \subseteq I} |a(\lambda)|^2 |I(\lambda)|^{p-1}.$$

Note that every function in $BMO^d(\mathbf{T})$ can be described by the Haar system, that is, an $L^2(\mathbf{T})$-function f belongs to $BMO^d(\mathbf{T})$ if and only if its Haar coefficients $\mathbf{a} = \{a(\lambda)\}$:

$$a(\lambda) = \langle f, h_\lambda \rangle = \int_{\mathbf{T}} f(t) h_\lambda(t) dt$$

satisfy

$$\sup_{I \in \mathcal{I}} T_{\mathbf{a},1}(I) < \infty. \tag{8.18}$$

See also [39]. Similarly, for $\mathcal{Q}_p^d(\mathbf{T})$ we have

Theorem 8.3.1. *Let $p \in (0,1)$. If $f \in \mathcal{Q}_p^d(\mathbf{T})$, then the sequence of its Haar coefficients $\mathbf{a} = \{a(\lambda)\}$ satisfies*

$$\sup_{I \in \mathcal{I}} T_{\mathbf{a},p}(I) < \infty. \tag{8.19}$$

Conversely, every sequence $\mathbf{a} = \{a(\lambda)\}$ satisfying (8.19) is the sequence of Haar coefficients of a unique $f \in \mathcal{Q}_p^d(\mathbf{T})$.

Proof. If $a(\lambda) = \langle f, h_\lambda \rangle$ and $I \in \mathcal{I}$, then

$$(f - f(I))1_I = \sum_{I(\lambda) \subseteq I} a(\lambda) h_\lambda$$

and thus

$$\Phi_f(I) = |I|^{-1} \sum_{I(\lambda) \subseteq I} |a(\lambda)|^2 = T_{\mathbf{a},1}(I).$$

It follows by the definition of $\Psi_{f,p}(I)$ and Lemma 8.3.1 that $\Psi_{f,p}(I) \approx T_{\mathbf{a},p}(I)$, and the result follows by Theorem 8.2.1.

This simple theorem suggests us to consider the wavelet bases. Recall that a wavelet is a function $\Psi \in L^2(\mathbf{R})$ such that the family of functions $\Psi_{j,k}(x) = 2^{j/2}\Psi(2^j x - k)$ where j and k range over $\mathbf{Z}$ (the set of all integers), is an orthonormal basis in $L^2(\mathbf{R})$. For such a family, let

$$\psi_{j,k}(x) = \sum_{l\in\mathbf{Z}} \Psi_{j,k}(x+l).$$

Then each $\psi_{j,k}$ is a function on $\mathbf{T}$ (i.e., a 1-periodic function on $\mathbf{R}$). Moreover, $\psi_{j,k}(x) = \psi_{j,k+2^j}(x)$ and $\psi_{j,k+1}(x) = \psi_{j,k}(x+2^{-j})$. In particular, there exists a Ψ so that $\{1\}\cup\{\psi_{j,k}\}$ ($j = 0,1,2,\cdots; k = 0,1,2,\cdots,2^j-1$) is a complete orthonormal basis in $L^2(\mathbf{T})$, viz., the 1-periodic wavelet basis. For convenience, we will write the shorter notation $\psi_{j,k}$ as ψ_λ, where $\lambda = (j,k)$. And for simplicity we consider only "good" wavelets, and thus suppose that each Ψ satisfies

$$\max\{|\Psi(x)|, |\Psi'(x)|\} \preceq (1+|x|)^{-2}, \quad x\in\mathbf{R};$$

but also Ψ has a compact support so that the support set of each ψ_λ obeys: $\mathrm{supp}\psi_\lambda \subseteq mI(\lambda)$, where m is a constant (fixed throughout the rest part of this section). For these, we refer to Meyer [92, Section 11 in Chapter 3] and Wojtaszczyk [130, Section 2.5].

Observe that the wavelet coefficients $\mathbf{b} = \{b(\lambda)\}$ of a $BMO(\mathbf{T})$-function are entirely controlled by $\sup_{I\in\mathcal{I}} T_{\mathbf{b},1}(I) < \infty$ [92, p.162]. This can be extended to $\mathcal{Q}_p(\mathbf{T})$ as follows.

Theorem 8.3.2. *Let $p\in(0,1)$. If $f\in\mathcal{Q}_p(\mathbf{T})$, then the sequence $\mathbf{b} = \{b(\lambda)\}$ of its wavelet coefficients:*

$$b(\lambda) = \langle f, \psi_\lambda\rangle = \int_{\mathbf{T}} f(x)\psi_\lambda(x)\,dx,$$

satisfies

$$\sup_{I\in\mathcal{I}} T_{\mathbf{b},p}(I) < \infty. \tag{8.20}$$

Conversely, every sequence $\mathbf{b} = \{b(\lambda)\}$ satisfying (8.20) is the sequence of wavelet coefficients of a unique $f\in\mathcal{Q}_p(\mathbf{T})$.

Proof. First, let $f\in\mathcal{Q}_p(\mathbf{T})$ and $I\in\mathcal{I}$. For $J\in\mathcal{I}_k(I)$, $k\in\mathbf{N}\cup\{0\}$, put

$$f = f_{mJ} + (f-f_{mJ})1_{mJ} + (f-f_{mJ})1_{\mathbf{T}\setminus mJ} = f_1+f_2+f_3.$$

Since $\mathrm{supp}\psi_\lambda\subseteq mI(\lambda)$, $\langle f_3,\psi_\lambda\rangle = 0$ if $I(\lambda)\subseteq J$. On the other hand, the integral of each wavelet ψ_λ is zero. So $\langle f,\psi_\lambda\rangle = \langle f_2,\psi_\lambda\rangle$, and, by (8.2) one has

$$\begin{aligned}\sum_{I(\lambda)\subseteq J} |\langle f,\psi_\lambda\rangle|^2 &\le \sum_\lambda |\langle f_2,\psi_\lambda\rangle|^2 = \|f_2\|_{L^2}^2 = |mJ|\Phi_f(mJ)\\ &\preceq \frac{1}{|mJ|}\int_{mJ}\int_{mJ}|f(x)-f(y)|^2\,dx\,dy.\end{aligned}$$

This gives that for $J\in\mathcal{I}_k(I)$,

$$T_{\mathbf{b},1}(J) = \frac{1}{|J|}\sum_{I(\lambda)\subseteq J}|b(\lambda)|^2 \le \frac{1}{|J||mJ|}\int_{mJ}\int_{mJ}|f(x)-f(y)|^2\,dx\,dy.$$

Using Lemma 8.3.1, we obtain in the same manner as for Lemma 8.1.2

$$\begin{aligned}
T_{\mathbf{b},p}(I) &\preceq \sum_{k=0}^{\infty} 2^{-pk} \sum_{J\in\mathcal{I}_k(I)} T_{\mathbf{b},1}(J) \\
&\preceq \int_{mI}\int_{mI} |f(x)-f(y)|^2 \sum_{k=0}^{\infty}\sum_{J\in\mathcal{I}_k(I)} \frac{1_{mJ}(x)1_{mJ}(y)}{2^{pk}|J|^2}\,dx\,dy \\
&\preceq |I|^{-p}\int_{mI}\int_{mI}\frac{|f(x)-f(y)|^2}{|x-y|^{2-p}}\,dx\,dy \\
&\preceq \|f\|^2_{\mathcal{Q}_{p,*}}.
\end{aligned}$$

Thus (8.20) follows.

Conversely, suppose that (8.20) holds; multiplying f by a constant, we may assume that $T_{\mathbf{b},p}(I) \le 1$ for every $I \in \mathcal{I}$. In particular, $T_{\mathbf{b},1}(I) \le T_{\mathbf{b},p}(I) \le 1$ for every $I \in \mathcal{I}$, and so

$$f = \sum_{\lambda} b(\lambda)\psi_\lambda \in BMO(\mathbf{T}),$$

with the sum converging e.g. in the weak* topology on $BMO(\mathbf{T})$. We will verify $f \in \mathcal{Q}_p(\mathbf{T})$.

Fix a (not necessarily dyadic) interval I of dyadic length and consider an interval $J \in \mathcal{I}_1(I)$. Let $\Lambda_0(J) = \{\lambda : mI(\lambda)\cap J \ne \emptyset\}$ and partition this set into

$$\begin{aligned}
\Lambda_1 = \Lambda_1(J) &= \{\lambda \in \Lambda_0(J) : |I(\lambda)| \le |J|\}, \\
\Lambda_2 = \Lambda_2(J) &= \{\lambda \in \Lambda_0(J) : |J| < |I(\lambda)| \le |I|\}, \\
\Lambda_3 = \Lambda_3(J) &= \{\lambda \in \Lambda_0(J) : |I| < |I(\lambda)|\}.
\end{aligned}$$

Since $\psi_\lambda = 0$ on J unless $\lambda \in \Lambda_0$ we have, on J, $f = f_1 + f_2 + f_3$, where

$$f_j = \sum_{\lambda\in\Lambda_j} b(\lambda)\psi_\lambda, \quad j = 1,2,3.$$

Hence, using the Cauchy-Schwarz inequality we get

$$\Phi_f(J) \le 3\big(\Phi_{f_1}(J) + \Phi_{f_2}(J) + \Phi_{f_3}(J)\big). \tag{8.21}$$

In what follows, we treat the three terms separately. First of all,

$$\Phi_{f_1}(J) \le |J|^{-1}\|f_1\|^2_{L^2} = |J|^{-1}\sum_{\lambda\in\Lambda_1}|b(\lambda)|^2. \tag{8.22}$$

Secondly, $|\nabla\psi_\lambda| \preceq |I(\lambda)|^{-3/2}$, and thus

$$|f_2(x) - f_2(y)| \preceq \sum_{\lambda\in\Lambda_2}|b(\lambda)||I(\lambda)|^{-3/2}|x-y|.$$

As a consequence, we have by letting $\epsilon = (1+p)/2$ and using the Cauchy-Schwarz inequality

$$\Phi_{f_2}(J) \preceq |J|^2\Big(\sum_{\lambda\in\Lambda_3}\frac{|b(\lambda)|}{|I(\lambda)|^{3/2}}\Big)^2$$
$$\preceq |J|^2\sum_{\lambda\in\Lambda_2}\frac{|b(\lambda)|^2}{|I(\lambda)|^3}\Big(\frac{|I(\lambda)|}{|J|}\Big)^\epsilon\sum_{\lambda\in\Lambda_2}\Big(\frac{|J|}{|I(\lambda)|}\Big)^\epsilon.$$

If $\lambda \in \Lambda_2$, then $I(\lambda)$ is a dyadic interval contained in an interval with the same center as J and length $(m+1)|I(\lambda)| + |J| \le (m+2)|J|$. Hence, for each $k \in \mathbf{N}$, there are at most $m+2$ such intervals $I(\lambda)$ with $|I(\lambda)| = 2^k|J|$. Moreover, there is a constant number of different λ for each such interval, and so the number of elements of $\{\lambda \in \Lambda_2 : |I(\lambda)| = 2^k|J|\}$ is finite for each $k \in \mathbf{N}$. Consequently,

$$\sum_{\lambda\in\Lambda_2}\Big(\frac{|J|}{|I(\lambda)|}\Big)^\epsilon \preceq \sum_{k=1}^{\infty}2^{-k\epsilon}\preceq 1$$

and

$$\Phi_{f_2}(J) \preceq \sum_{\lambda\in\Lambda_2}|b(\lambda)|^2\Big(\frac{|J|}{|I(\lambda)|}\Big)^{2-\epsilon}|I(\lambda)|^{-1}. \tag{8.23}$$

Thirdly, we similarly have

$$|f_3(x)-f_3(y)| \preceq \sum_{\lambda\in\Lambda_3}\frac{|x-y|}{|I(\lambda)|^{3/2}} \preceq |x-y|\sum_{\lambda\in\Lambda_3}|I(\lambda)|^{-1},$$

by

$$|b(\lambda)||I(\lambda)|^{-1/2} \le T^{1/2}_{\mathbf{b},p}(I(\lambda)) \le 1.$$

Again, there is a bounded number of terms for each $I(\lambda)$, and now $|I(\lambda)| = 2^k|I|$, $k \in \mathbf{N}$; hence $|f_3(x) - f_3(y)| \preceq |x-y||I|^{-1}$ and

$$\Phi_{f_3}(J) \preceq |J|^2|I|^{-2}. \tag{8.24}$$

Consequently, by the above estimates: (8.21) through (8.24),

$$\Phi_f(J) \preceq \frac{1}{|J|}\sum_{\lambda\in\Lambda_1}|b(\lambda)|^2 + \sum_{\lambda\in\Lambda_2}\frac{|b(\lambda)|^2}{|I(\lambda)|}\Big(\frac{|J|}{|I(\lambda)|}\Big)^{2-\epsilon} + \Big(\frac{|J|}{|I|}\Big)^2.$$

Summing over $J \in \mathcal{I}_1(I)$ we obtain

$$\begin{aligned}\Psi_{f,p}(I) &= \sum_{J\in\mathcal{I}_1(I)}\Big(\frac{|J|}{|I|}\Big)^p\Phi_f(J)\\ &\preceq \sum_{J\in\mathcal{I}_1(I)}\sum_{\lambda\in\Lambda_1(J)}\frac{|b(\lambda)|^2}{|J|}\Big(\frac{|J|}{|I|}\Big)^p\\ &\quad+\sum_{J\in\mathcal{I}_1(I)}\sum_{\lambda\in\Lambda_2(J)}|b(\lambda)|^2|J|^{p+2-\epsilon}|I(\lambda)|^{\epsilon-3}|I|^{-p}\\ &\quad+\sum_{J\in\mathcal{I}_1(I)}\Big(\frac{|J|}{|I|}\Big)^{p+2}.\end{aligned} \tag{8.25}$$

The final sum equals

$$\sum_{j=0}^{\infty} 2^j (2^{-j})^{p+2} = \sum_{j=0}^{\infty} 2^{-(1+p)j} \preceq 1.$$

In the two double sums, we interchange the order of summation. If λ occurs there, then $|I(\lambda)| \leq |I|$ and $mI(\lambda) \cap I \neq \emptyset$; thus, if we let $\mathcal{J}(I)$ be the set of dyadic intervals J of the same size as I with $mJ \cap I \neq \emptyset$, it follows that $I(\lambda) \in \mathcal{I}_1(J)$ for some $J \in \mathcal{J}(I)$.

Fix such a λ, with $|I(\lambda)| = 2^{-k}|I|$. For each $j \leq k$, there are at most finite many intervals $J \in \mathcal{I}_j(I)$ with $\lambda \in \Lambda_1(J)$, each contributing $2^{(1-p)j}|I|^{-1}|b(\lambda)|^2$ to the first double sum in (8.25). Similarly, for each integer $j > k$, there are at most $C2^{(j-k)}$ (where C is an absolute constant) intervals $J \in \mathcal{I}_j(I)$ with $\lambda \in \Lambda_2(J)$, each contributing

$$2^{-j(p+2-\epsilon)-k(\epsilon-3)}|I|^{-1}|b(\lambda)|^2$$

to the second double sum in (8.25). These get together to yield at most

$$C|I|^{-1}|b(\lambda)|^2 \left(\sum_{j=0}^{k} 2^{(1-p)j} + \sum_{j=k+1}^{\infty} \frac{2^{(1-p)k}}{2^{(1+p-\epsilon)}} \right) \preceq 2^{(1-p)k} \frac{|b(\lambda)|^2}{|I|}.$$

As a result, (8.25) gives

$$\begin{aligned} \Psi_{f,p}(I) &\preceq \sum_{J\in\mathcal{J}(I)} \sum_{k=0}^{\infty} \sum_{I(\lambda)\in\mathcal{I}_k(J)} 2^{(1-p)k}|I|^{-1}|b(\lambda)|^2 + 1 \\ &\preceq \sum_{J\in\mathcal{J}(I)} T_{\mathbf{b},p}(I') + 1. \end{aligned}$$

We have proved that $\Psi_{f,p}(I) \preceq 1$ for every interval I of dyadic length. Since the same estimate applies to every translate $I + t$, Lemma 8.1.3 shows that

$$\int_I \int_I \frac{|f(x) - f(y)|^2}{|x-y|^{2-p}} dxdy \preceq |I|^p$$

is valid for every interval I of dyadic length. Therefore, Corollary 8.1.1 implies $f \in \mathcal{Q}_p(\mathbf{T})$.

Uniqueness of f follows from the uniqueness in $BMO(\mathbf{T})$; if $f, g \in \mathcal{Q}_p(\mathbf{T}) \subseteq BMO(\mathbf{T})$ have the same wavelet coefficients, then they define the same linear functional on the predual space of $BMO(\mathbf{T})$ and thus $f = g$ as elements of $BMO(\mathbf{T})$ (i.e. modulo constants), see [92, Section 5.6] once again.

Corollary 8.3.1. *Let $p \in (0,1)$. Then $U : \sum b(\lambda)\psi_\lambda \to \sum a(\lambda)h_\lambda$ sets up an isomorphism between $\mathcal{Q}_p(\mathbf{T})$ and $\mathcal{Q}_p^d(\mathbf{T})$ with the sums interpreted formally or as converging in suitable weak topologies.*

Notes

8.1 Section 8.1 is one of the main topics of Janson's paper [80]. Corollary 8.1.1 tells us that restriction of $\mathcal{Q}_p(\mathbf{T})$ to dyadic intervals would give $\mathcal{Q}_p^d(\mathbf{T})$.

8.2 Section 8.2 is also from [80]. Theorem 8.2.2 is similar to the one for $BMO(\mathbf{T})$ [64]. Of course, Theorem 8.2.3 reveals a close relation between $\mathcal{Q}_p(\mathbf{T})$ and $BMO(\mathbf{T})$. Note that Theorem 8.2.6 is an extension to $\mathcal{Q}_p(\mathbf{T})$ of a result by Garnett and Jones [67] for $BMO(\mathbf{T})$. Here, it is worth pointing out that the space $L^q(\mathcal{Q}_p^d)$ is not the usual Lebesgue space of Banach space valued functions, defined as the closure of simple functions in the obvious norm. The problem is that $\mathcal{Q}_p^d(\mathbf{T})$ is not separable, and it is easily seen that if e.g. $F(t,x) = 1_{\{(t,x):0<x<t\}}$, then F belongs to $L^q(\mathcal{Q}_p^d)$ for any $q \in (0,\infty]$, however $\|F(t,\cdot) - F(s,\cdot)\|_{\mathcal{Q}_p^d} \geq 1/4$ for a.e. s and t, and so there is no separable subspace of $\mathcal{Q}_p^d(\mathbf{T})$ that contains $F(t,\cdot)$ for a.e. t, as required by the standard definition, see e.g. Dunford-Schwartz [49].

8.3 The results of Section 8.3 are essentially from the paper [60] by Essén, Janson, Peng and Xiao, in which their major aim is to study the Q classes in $\mathbf{R}^n$ (whose new progress has been made in Dafni-Xiao [45]). For another approach to deal with the wavelet coefficients of functions in $\mathcal{Q}_p(\mathbf{T})$, see also Xiao [136]. According to the above definition of a wavelet, the Haar function is a wavelet. However, we should observe that the Haar function is very well localized. The supports of the induced functions $h_{j,k}$ are dyadic intervals and are easily understood. The major problem is that the Haar function is not continuous. For this reason, Carleson introduced a Lip1 function ϕ with support in $(-\delta, 1+\delta)$ (for some small constant $\delta > 0$) and defined a family of functions $\{\phi_\omega\}$ via translating and scaling ϕ to the dyadic interval ω, as well as normalizing ϕ_ω with L^2-norm 1. Although $\{\phi_\omega\}$ is not of the orthonormal property, he proved that there is a function ϕ such that $f \in BMO(\mathbf{T})$ if and only if $f = \sum_{\omega\in\mathcal{I}} c(\omega)\phi_\omega$, with convergence in $L^2(\mathbf{T})$, where the coefficients $\mathbf{c} = \{c(\omega)\}_{\omega\in\mathcal{I}}$ obey $\sup_{I\in\mathcal{I}} T_{\mathbf{c},1}(I) < \infty$. See Carleson's work [39] for further details. Naturally, this result can be generalized to $\mathcal{Q}_p(\mathbf{T})$ through Theorem 8.3.2.

8.4 A modification of the Haar function may derive a characterization of the extreme points of the closed unit ball of $\mathcal{Q}_p(\mathbf{T})$. See Wirths-Xiao [128] for an account.

References

1. Adams D.R. (1988) A note on Choquet integral with respect to Hausdorff capacity. In "Function Spaces and Applications," Lund 1986, Lecture Notes Math. 1302: 115-124
2. Ahern P.R. (1979) The mean modulus and the derivative of an inner function. Indiana Univ. Math. J. 28: 311-347
3. Ahern P., Jevtić M. (1990) Inner multipliers of the Besov space, $0 < p \leq 1$. Rocky Mountain J. Math. 20: 753-764
4. Ahlfors L.V. (1973) Conformal invariants. McGraw-Hill, New York
5. Aleman A. (1992) Hilbert spaces of analytic functions between the Hardy and the Dirichlet spaces. Proc. Amer. Math. Soc. 115: 97-104
6. Anderson J.M. (1979) On division by inner functions. Comm. Math. Helv. 54: 309-317
7. Arcozzi N., Sawyer E., Rochberg R. (2000) Carleson measures for analytic Besov spaces. Preprint
8. Anderson J.M., Clunie, Pommerenke Ch. (1974) On Bloch functions and normal functions. J. Reine Angew. Math. 270: 12-37
9. Andersson M., Carlsson H. (2000) Q_p spaces in strictly pseudoconvex domains. Preprint
10. Arazy J., Fisher S. (1984) Some aspects of the minimal, Möbius invariant space of analytic functions in the unit disc. Lecture Notes Math. 1070: 24-44
11. Arazy J., Fisher S., Janson S., Peetre J. (1990) An identity for reproducing kernels in a planar domain and Hilbert-Schmidt Hankel operators. J. Reine Angew. Math. 406: 179-199
12. Arazy J., Fisher S., Peetre J. (1985) Möbius invariant function spaces. J. Reine Angew. Math. 363: 110-145
13. Aulaskari R. (2000) On Q_p functions. Complex analysis and related topics (Cuernavaca, 1996). Oper. Theory Adv. Appl. (Birkhuser, Basel) 114: 21-29
14. Aulaskari R., Chen H. (1998) On $Q_p(R)$ and $Q_p^{\#}(R)$ for Riemann surfaces . Preprint
15. Aulaskari R., Csordas D. (1995) Besov spaces and $Q_{p,0}$ classes. Acta Sci. Math. (Szeged) 60: 31-48
16. Aulaskari, R., Danikas N., Zhao R. (1999) The algebra property of the integrals of some analytic functions in the unit disk. Ann. Acad. Sci. Fenn. A I Math. 24: 343-351
17. Aulaskari R., Girela D., Wulan H. (1998) Q_p spaces, Hadamard products and Carleson measures. Preprint
18. Aulaskari R., Girela D., Wulan H. (2001) Taylor coefficients and mean growth of the derivative of Q_p functions. J. Math. Anal. Appl. 258: 415-428
19. Aulaskari R., He Y., Ristioja J., Zhao R. (1998) Q_p spaces on Riemann surfaces. Canad. J. Math. 50: 449-464

20. Aulaskari R., Lappan P. (1994) Criteria for an analytic function to be Bloch and a harmonic or meromorphic function to be normal. Complex analysis and its applications, Pitman Res. Notes Math. 305, Longman Sci. Tech., Harlow: 136-146
21. Aulaskari R., Lappan P., Xiao J., Zhao R. (1997) On α-Bloch spaces and multipliers of Dirichlet spaces, J. Math. Anal. Appl. 209: 103-121
22. Aulaskari R., Nowak M., Zhao R. (1998) The nth derivative characterization of the Möbius bounded Dirichlet space. Bull. Austral. Math. Soc. 58: 43-56
23. Aulaskari R., Perez-Gonzalez F., Wulan H. (2000) Some inequalities for Q_p functions. Preprint
24. Aulaskari R., Stegenga D., Xiao J. (1996) Some subclasses of BMOA and their characterization in terms of Carleson measures. Rocky Mountain J. Math. 26: 485-506
25. Aulaskari R., Stegenga D., Zhao R. (1996) Random power series and Q_p. XVIth Rolf Nevanlinna Colloquium, Eds.:Laine/Martio, Water de Gruyter Co. Berlin New York, pp. 247-255
26. Aulaskari R., Wulan H., Zhao R. (2000) Carleson measures and some classes of meromorphic functions. Proc. Amer. Math. Soc. 28: 2329-2335
27. Aulaskari R., Xiao J., Zhao R. (1995) On subspaces and subsets of BMOA and UBC. Analysis 15: 101-121
28. Aulaskari R., Zhao R. (1999) Boundedness and compactness properties of the Libera transform. Complex analysis and differential equations (Uppsala, 1997), Acta Univ. Upsaliensis Skr. Uppsala Univ. C Organ. Hist. 64: 69-80
29. Axler S. (1986) The Bergman space, the Bloch space and commutators of multiplication operators. Duke Math. J. 53: 315-332.
30. Baernstein II A. (1980) Analytic functions of bounded mean oscillation. Aspects of Contemporary Complex Analysis, Academic Press, pp. 3-36
31. Benke G., Chang D.C. (2000) A note on weighted Bergman spaces and the Cesàro operator. Nagoya Math. J. 159: 25-43
32. Bennett G., Stegenga D.A., Timoney R. (1981) Coefficients of Bloch and Lipschitz functions. Illinois J. Math. 25: 520-531
33. Bergh J. (1988) Functions of bounded mean oscillation and Hausdorff-Young type theorems. Lecture Notes in Math. 1302: 130-136
34. Bourdon P.S., Cima J.A., Matheson A.L. (1999) Compact composition operators on $BMOA$. Trans. Amer. Math. Soc. 351: 2183-2196
35. Bourdon P.S., Shapiro J., Sledd W. (1989) Fourier series, mean Lipschitz spaces, and bounded mean oscillation. Analysis at Urbana, Vol. I (Urbana, IL, 1986-1987), London Math. Soc. LNS 137: 81-110
36. Brown L., Shields A.L. (1991) Multipliers and cyclic vectors in the Bloch space. Michigan Math. J. 38: 141-146
37. Carleson L. (1950) On a class of meromorphic functions and its associated exceptional sets. Thesis, Uppsala Univ.
38. Carleson L. (1962) Interpolations by bounded analytic functions and the corona problem. Ann. Math. 76: 547-559
39. Carleson L. (1980) An explicit unconditional basis in H^1. Bull. Sci. Math., 2ᶜ série 104: 405-416
40. Cnops J., Delanghe R. (1999) Möbius invariant spaces in the unit ball. In Begehr special issue, Appl. Anal. 73: 45-64
41. Cnops J., Delanghe R., Gürlebeck K., Shapiro M.V. (1998) Q_p-spaces in Clifford analysis. In Proc. Conf. Dirac Oper. Italy, Oct., 1998
42. Cohran W.G., Shapiro J.H., Ullrich D.C. (1993) Random Dirichlet functions: multipliers and smoothness. Canad. J. Math. 45: 255-268
43. Cowen C., MacCluer B. (1995) Composition operators on spaces of analytic functions. CRC Press, Boca Raton

44. Cuerva G.J., Rubio de Francia J.L. (1985) Weighted norm inequalities and related topics. North-Holland, Amsterdam
45. G. Dafni, Xiao J. (2001) Affine invariant arising from square potential spaces. Preprint
46. Danikas N., Mouratides C. (2000) Blaschke products in Q_p spaces. Complex Variables Theory Appl. 43: 199-209
47. Danikas N., Ruscheweyh S., Siskakis A. (1994) Metrical and topological properties of a generalized Libera transform. Arch. Math. (Basel) 63: 517-524
48. Danikas N., Siskakis A. (1993) The Cesàro opeator on bounded analytic functions. Analysis 13: 295-299
49. Dunford N., Schwartz J.T. (1958) Linear operators, Part I. Interscience, New York
50. Duren P. (1970) Theory of H^p spaces. Academic Press
51. Duren P. (1985) Random series and bounded mean oscillation. Michigan Math. J. 32: 81-86
52. Dyakonov K.M. (1997) Factorization of smooth analytic functions via Hilbert-Schmidt operators. St. Petersburg Math. J. 8: 543-569
53. Dyakonov K.M. (1997) Equivalent norms on Lipschitz-type spaces of holomorphic functions. Acta Math. 178: 143-167
54. Dyakonov K.M. (1998) Besov spaces and outer functions. Michigan Math. J. 45: 143-157
55. Dyakonov K.M. (1993) Division and multiplication by inner functions and embedding theorems for star-invariant subspaces. Amer. J. Math. 115: 881-992
56. Dyakonov K.M., Girela D. (2000) On Q_p spaces and preudoanalytic extension. Ann. Acad. Sci. Fenn. Ser. A I Math. 25: 477-486
57. Dyn'kin E.M. (1993) The pseudoanalytic extension. J. Anal. Math. 60: 45-70
58. Earl J.P. (1970) On the interpolation of bounded sequences by bounded functions. J. London Math. Soc.(2) 2: 544-548
59. Essén M. (2001) Q_p spaces. Univ. Joensuu Dept. Math. Rep. Ser.4: 9-40
60. Essén M., Janson S., Peng L., Xiao J. (2000) Q spaces of several real variables. Indiana Univ. Math. J. 49: 575-615
61. Essén M., Wulan H. (2000) Carleson type measures and their applications. Complex Variables Theory Appl. 42: 67-88
62. Essén M., Xiao J. (1997) Some results on Q_p spaces, $0 < p < 1$. J. Reine Angew. Math. 485: 173-195
63. Essén M., Xiao J. (2001) Q_p spaces – a survey. Univ. Joensuu Dept. Math. Rep. Ser. 4: 41-60
64. Fefferman C., Stein E.M. (1972) H^p spaces of several variables. Acta Math. 129: 137-193
65. Frazier M., Jawerth B., Weiss G. (1991) Littlewood-Paley theory and the study of function spaces. CBMS Regional Conference Series in Math. 79
66. Garnett J. (1981) Bounded analytic functions. Academic Press, New York
67. Garnett J., Jones P. (1982) BMO from dyadic BMO. Pacific J. Math. 99: 351-371
68. Gauthier P., Xiao J. (1999) Functions of bounded expansion: normal and Bloch functions. J. Austral. Math. Soc. (Ser A) 66: 168-188
69. Gauthier P., Xiao J. (2000) BiBloch type maps: existence and beyond. Preprint.
70. Girela D. (2001) Analytic functions of bounded mean oscillation. Complex function spaces (Mekrijrvi, 1999), Univ. Joensuu Dept. Math. Rep. Ser. 4: 61-171
71. Girela D., Marquez M. (1999) Some remarks on Carleson measures and Q_p spaces. In: Proc. Symposium on Complex Analysis and Differential Equations, June 15-18, 1997. Acta Univ. Upsaliensis Skr. Uppsala Univ. C Organ. Hist. 64: 169-178
72. Gotoh Y. (1999) On uniform and relative uniform domains. Preprint
73. Gürlebeck K., Kähler U., Shapiro M.V., Tovar L.M. (1999) On Q_p-spaces of quaternion-valued functions. Complex Variables 39: 115-135

74. Havin V.P. (1971) On the factorization of analytic functions smooth up to the boundary. Zap. Nauch. Sem. LOMI 22: 202-205
75. Holland F., Twomey J.B. (1985) Explicit examples of Bloch functions in every H^p space, but not in $BMOA$. Proc. Amer. Math. Soc. 95: 227-229
76. Holland F., Walsh D. (1984) Boundedness criteria for Hankel operators. Proc. Royal Irish Acad. Sect. A 84: 141-154
77. Holland F., Walsh D. (1986) Criteria for membership of Bloch space and its subspaces, BMOA. Math. Ann. 273: 317-335
78. Hörmander L. (1967) Generators for some rings of analytic functions. Bull. Amer. Math. Soc. 73: 943-949
79. Hu P., Shi J., Zhang W. (1999) The Möbius boundedness of the space Q_p. J. Austral Math. Soc. (Ser. A) 66: 373-378
80. Janson S. (1999) On the space Q_p and its dyadic counterpart. In: Proc. Symposium on Complex Analysis and Differential Equations, June 15-18, 1997. Acta Univ. Upsaliensis Skr. Uppsala Univ. C Organ. Hist. 64: 194-205
81. Jafari F. et al. (1998), Studies on Composition Operators, Contemp. Math. 213
82. John F., Nirenberg L. (1965) On functions of bounded mean oscillation. Comm. Pure Appl. Math. 18: 415-426
83. Jones P. (1983) L^∞-estimates for the $\bar{\partial}$ problem in a half plane. Acta Math. 150: 137-152
84. Kerman R., Sawyer E. (1988) Carleson measures and multipliers of Dirichlet-type spaces. Trans. Amer. Math. Soc. 309: 87-98
85. Latvala V. (1999) On subclasses of $BMO(B)$ for solutions of elliptic equations. Analysis 19: 103-116
86. Lou Z. J. (2001) Composition operators on Q^p spaces. J. Austral. Math. Soc. (Ser. A) 70: 161-188
87. Lindström M, Makhmutov S., Taskinen J. (2001) The essential norm of Bloch-to-Q_p composition operator. Preprint
88. Madigan K. (1993) Composition operators into Lipschitz type spaces. Thesis, SUNY Albany
89. Madigan K., Matheson A.L. (1995) Compact composition operators on the Bloch space. Trans. Amer. Math. Soc. 347: 2679-2687
90. Mateljevic M., Pavlovic M. (1983) L^p-behaviour of power series with positive coefficients and Hardy spaces. Proc. Amer. Math. Soc. 87: 309-316
91. Metzger T.A. (1981) On BMOA for Riemann surfaces. Canad. J. Math. 33: 1255-1260
92. Meyer Y. (1992), Wavelets and Operators. Cambridge Univ. Press
93. Montes-Rodriguez A. (1999), The essential norm of a composition operator on Bloch spaces. Pacific J. Math. 118: 339-351
94. Nicolau A. (1990) The corona property for bounded functions in some Besov spaces. Proc. Amer. Math. Soc. 110: 135-140
95. Nicolau A., Xiao J. (1997) Bounded functions in Möbius invariant Dirichlet spaces. J. Funct. Anal. 150: 383-425
96. Ortega J.M., Fabrega J. (1995) Pointwise multipliers and corona type decompositions in $BMOA$. Ann. Inst. Fourier Grenoble 46: 1-26.
97. Ortega J.M., Fabrega J. (1996) The corona type decomposition in some Besov spaces. Math. Scand. 78: 93-111
98. Ouyang C., Yang W., Zhao R. (1982) Möbius invariant Q_p spaces associated with the Green's function on the unit ball of C^n. Pacific J. Math. 182: 68-100
99. Petersen K.E. (1977) Brownian motion, Hardy spaces and bounded mean oscillation. London Math. Soc. LNS 28
100. Pommerenke Ch. (1970) On Bloch functions. J. London Math. Soc. (2)2: 689-695
101. Pommerenke Ch. (1977) Schlichte funktionen und analytische funktionen von beschränkten mittlerer oszillation. Comm. Math. Helv. 52: 591-602

102. Rabindranathan M. (1972) Toeplitz operators and division by inner functions. Indiana Univ. Math. J. 22: 523-529
103. Ramey W., Ullrich D. (1991) Bounded mean oscillation of Bloch pull-backs. Math. Ann. 291: 591-606
104. Resendis L.F., Tovar L.M. (1999) Carleson measures, Blaschke products and Q_p-spaces. In: Proc. Symposium on Complex Analysis and Differential Equations, June 15-18, 1997. Acta Univ. Upsaliensis Skr. Uppsala Univ. C Organ. Hist. 64: 296-305
105. Rochberg R., Wu Z. (1993) A new charaterization of Dirichlet type spaces and applications. Illinois J. Math. 37: 101-122
106. Rubel L., Timoney R. (1979) An extremal property of the Bloch space. Proc. Amer. Math. Soc. 75: 45-49
107. Rudin W. (1955) The radial variation of analytic functions. Duke Math. J. 22: 235-242
108. Rudin W. (1974) Real and complex analysis. 2nd ed., MaGraw-Hill, New York
109. Sadosky C. (1979) Interpolation of Operators and Singular Integrals. Marcel Dekker, INC, New York and Basel
110. Shapiro J.H. (1993) Composition Operators and Classical Function Theory. Springer-Verlag, New York
111. Siskakis A.G. (1998) Semigroups of composition operators on spaces of analytic functions, a review. Contemp. Math. 213: 229-252
112. Sledd W.T. (1981) Random series which are BMO or Bloch. Michigan Math. J. 28: 259-266
113. Sledd W.T., Stegenga D. (1981) An H^1 multiplier theorem. Ark. Mat. 19: 265-270
114. Smith W. (1999) Compactness of composition operators on $BMOA$. Proc. Amer. Math. Soc. 127: 2715-2726
115. Smith W., Zhao R. (1997) Composition operators mapping into the Q_p spaces. Analysis 17: 239-263
116. Stegenga D. (1973) Bounded Toeplitz operators on H^1 and applications of the duality between H^1 and the functions of bounded mean oscillation. Amer. J. Math. 98: 573-589
117. Stegenga D. (1980) Multipliers of the Dirichlet space. Illinois J. Math. 24: 113-139
118. Stein E.M. (1970) Singular integrals and differentiability properties of functions. Princeton Univ. Press, Princeton
119. Stein E.M. (1993) Harmonic analysis, Real-variable methods, orthogonality and oscillatory integrals. Princeton Univ. Series, 43, Princeton Univ. Press, Princeton
120. Strichartz R.S. (1980) Bounded mean oscillation and Sobolev spaces. Indiana Univ. Math. J. 29: 538-558
121. Stroethoff K. (1989) Besov-type characterizations for the Bloch space. Bull. Austral. Math. Soc. 39: 405-420
122. Stroethoff K. (1990) Nevanlinna-type characterizations for the Bloch space and related spaces. Proc. Edinburgh Math. Soc. 33: 123-141
123. Suárez D. (2001) Meromorphic and harmonic functions inducing continuous maps from M_{H^∞} into the Riemann sphere. J. Funct. Anal. 183: 164-210
124. Tjani M. (1996) Compact composition operators on some Möbius invariant Banach spaces. Ph.D. Thesis, Michigan State Univ.
125. Verbitskii I.E. (1985) Multipliers in spaces with "fractional" norms and inner functions (in Russian). Sibirsk Mat. Zh. 26(2): 51-72
126. Wirths K.J., Xiao J. (1996) Image areas of functions in the Dirichlet type spaces and their Möbius invariant subspaces. Ann. Univ. Mariae Curie-Sklodowska Sect. A 50: 241-247
127. Wirths K.J., Xiao J. (2001) Recognizing $Q_{p,0}$ functions per Dirichlet space structure. Bull. Belg. Math. Soc. 8: 47-59
128. Wirths K.J., Xiao J. (2000) Extreme points in vanishing $Q_p(\partial\Delta)$-space. Preprint

129. Wirths K.J., Xiao J. (2000) Global integral criteria for composition operators. Preprint
130. Wojtaszczyk R. (1997) A Mathematical Introduction to Wavelets, Cambridge Univ. Press
131. Wulan H. (1998) On some classes of meromorphic functions. Ann. Acad. Sci. Fenn. Math. Diss. 116
132. Wulan H., Wu P. (2001) Characterization of Q_T spaces. J. Math. Anal. Appl. 254: 484-497
133. Xiao J. (1994) Carleson measure, atomic decomposition and free interpolation from Bloch space. Ann. Acad. Sci. Fenn. Ser. A I Math. 19:35-46
134. Xiao J. (1999) Outer functions in Q_p and $Q_{p,0}$. Preprint
135. Xiao J. (2000) The Q_p corona theorem. Pacific J. Math. 194: 491-509
136. Xiao J. (2000) Some essential properties of $Q_p(\partial\Delta)$-spaces. J. Fourier Anal. Appl. 6: 311-323
137. Xiao J. (2000) Composition operators: $\mathcal{N}_\alpha$ to the Bloch space to $\mathcal{Q}_\beta$. Studia Math. 139: 245-260
138. Xiao J. (2000) Biholomorphically invariant families amongst Carleson class. Preprint
139. Xiao J. (2001) Composition operators associated with Bloch-type spaces. Complex Variables Theory Appl.: to appear
140. Yamashita Y. (1980) Gap series and α-Bloch functions. Yokohama Math. J. 28: 31-36.
141. Yang W. (1998) Carleson type measure characterization of Q_p spaces. Analysis 18: 345-349
142. Yang W. (1999) Vanishing Carleson type measure characterization of $Q_{p,0}$. C. R. Math. Acad. Sci. R. Can. 21: 1-5
143. Zhao R. (1996) On a general family of function spaces. Ann. Acad. Sci. Fenn. Math. Diss. 105
144. Zhu K. (1990) Operator theory in function spaces. Pure and Applied Math., Marcel Dekker, New York
145. Zorboska N. (1998) Composition operators on weighted Dirichlet spaces. Proc. Amer. Math. Soc. 126: 2013-2023
146. Zygmund A. (1959) Trigonometric series, Vol I and II. Cambridge Univ. Press

Index

4. Lecture Notes are printed by photo-offset from the master-copy delivered in camera-ready form by the authors. Springer-Verlag provides technical instructions for the preparation of manuscripts. Macro packages in TEX, LATEX2e, LATEX2.09 are available from Springer's web-pages at

http://www.springer.de/math/authors/b-tex.html.

Careful preparation of the manuscripts will help keep production time short and ensure satisfactory appearance of the finished book.

The actual production of a Lecture Notes volume takes approximately 12 weeks.

5. Authors receive a total of 50 free copies of their volume, but no royalties. They are entitled to a discount of 33.3 % on the price of Springer books purchase for their personal use, if ordering directly from Springer-Verlag.

Commitment to publish is made by letter of intent rather than by signing a formal contract. Springer-Verlag secures the copyright for each volume. Authors are free to reuse material contained in their LNM volumes in later publications: A brief written (or e-mail) request for formal permission is sufficient.

Addresses:

Professor J.-M. Morel
CMLA, Ecole Normale Supérieure de Cachan
61 Avenue du Président Wilson
94235 Cachan Cedex France
E-mail: Jean-Michel.Morel@cmla.ens-cachan.fr

Professor B. Teissier
Université Paris 7
UFR de Mathématiques
Equipe Géométrie et Dynamique
Case 7012
2 place Jussieu
75251 Paris Cedex 05
E-mail: Teissier@ens.fr

Professor F. Takens, Mathematisch Instituut,
Rijksuniversiteit Groningen, Postbus 800,
9700 AV Groningen, The Netherlands
E-mail: F.Takens@math.rug.nl

Springer-Verlag, Mathematics Editorial, Tiergartenstr. 17
D-69121 Heidelberg, Germany
Tel.: *49 (6221) 487-701
Fax: *49 (6221) 487-355
E-mail: lnm@Springer.de

9 783540 426257